ROYAL HISTORICAL SOCIETY

STUDIES IN HISTORY

New Series

WOMEN'S BODIES AND DANGEROUS TRADES IN ENGLAND 1880–1914

Studies in History New Series

Editorial Board

Professor David Eastwood (*Convenor*)
Professor Michael Braddick
Dr Steven Gunn
Dr Janet Hunter (*Economic History Society*)
Professor Aled Jones (*Literary Director*)
Professor Colin Jones
Professor Mark Mazower
Professor Miles Taylor
Dr Simon Walker
Professor Julian Hoppit (*Honorary Treasurer*)

This series is supported by an annual subvention from the Economic History Society

WOMEN'S BODIES AND DANGEROUS TRADES IN ENGLAND 1880–1914

Carolyn Malone

THE ROYAL HISTORICAL SOCIETY
THE BOYDELL PRESS

© Carolyn Malone 2003

All Rights Reserved. Except as permitted under current legislation
no part of this work may be photocopied, stored in a retrieval system,
published, performed in public, adapted, broadcast,
transmitted, recorded or reproduced in any form or by any means,
without the prior permission of the copyright owner

First published 2003

A Royal Historical Society publication
Published by The Boydell Press
an imprint of Boydell & Brewer Ltd
PO Box 9, Woodbridge, Suffolk IP12 3DF, UK
and of Boydell & Brewer Inc.
PO Box 41026, Rochester, NY 14604–4126, USA
website: www.boydell.co.uk

ISBN 0 86193 264 1

ISSN 0269–2244

A catalogue record for this book is available
from the British Library

Library of Congress Catalog Card Number: 2003007950

This book is printed on acid-free paper

Printed in Great Britain by
St Edmundsbury Press Ltd, Bury St Edmunds, Suffolk

Contents

		Page
List of tables		vi
Acknowledgements		ix
Abbreviations		xi
Introduction		1
1	'Discovering' the dangers of women's work, 1830–1880s	9
2	Nails, chains and reproduction	20
3	Dangerous trades regulations and the white lead trade	33
4	Dangerous trades regulations and the pottery trade	52
5	Narratives of bodily danger: the new journalistic press	74
6	Medical men, sexual science and dangerous trades regulations	95
7	Feminists, dangerous trades and the state	120
Epilogue		139
Bibliography		151
Index		165

List of Tables

1. Illness among pottery workers, by sex — 57
2. Lead poisoning in the white lead trade, 1898 — 101
3. Lead poisoning: in-patients, Royal Infirmary, Newcastle-upon-Tyne, 1892–1900 — 102
4. Lead poisoning in the pottery and white lead trades, by sex, 1899 — 103
5. Lead poisoning in the white lead trade, by sex, 1900–14 — 116
6. Lead poisoning in the white lead trade, by sex, 1900–9 — 117
7. Lead poisoning in the pottery trade, by sex, 1900–9 — 118

FOR ROB

Acknowledgements

Since its first incarnation as a PhD dissertation at the University of Rochester, this work has undergone a series of transformations. Stewart Weaver and Bonnie Smith read the dissertation with great care and made valuable suggestions for its improvement. My greatest debt is indeed to Bonnie Smith, who introduced me to the exciting field of women's history and challenged me to expand my intellectual boundaries. Over the years her invaluable criticism has prompted me to reconceptualise this project.

Brigit Collins at the Trades Union Congress Library was especially helpful. I also owe a great deal to the skill and energy of the staff at the British Library and at its Newspaper Library at Colindale, at the British Library of Political and Economic Sciences, the Labour Party Archives, the Public Record Office at Kew and at the Wellcome Institute for the History of Medicine.

Grants from Georgia Southern University and Ball State University have helped support research trips to England. My selection as Ball State's Westminster Scholar provided a leave that was crucial to the completion of this project. I have had the opportunity to work with many exceptional colleagues at those institutions, especially Melody Alexander, Anne Bailey, John Barber, Jim Connolly, Tony Edmonds, Anastatia Sims, Chris Thompson and Phyllis Zimmerman. I appreciate their friendship and support.

I want to thank Colin Jones of the Royal Historical Society for his assistance. The anonymous readers provided valuable suggestions that have strengthened this book. I am especially grateful for Christine Linehan's editorial expertise and patience with a trans-Atlantic author.

Chapters 1, 3 and 5 of this book are based substantially on articles published in the *Journal of British Studies* lxxiii (1998), the *Journal of Women's History* viii (1996), and *Albion* xxxi (1999). I want to thank the editors of those journals, the North American Conference on British Studies, and the University of Chicago Press and Indiana University Press for permission to reproduce them here.

Family and friends have provided such crucial encouragement and support over the years. Elaine Bailey has been the best friend anyone could ask for, sharing the trials, tribulations and joys of life for two decades. I would like to thank the Halls, especially Bob and Helen who have always appreciated the life of the mind. Bob, Sandy and Eddie Lattanzio and Jimmy Malone have provided a fun and lively family life. I owe a great debt to my parents, Bill and Barbara, who have always encouraged me to pursue my intellectual interests. They have shown their love and support in so many ways including visits to England during my numerous research trips. My father (and brother) forced me out of the archives and into the pub while my mother indulged me at

Liberty's and the Ritz. Finally, I want to thank Robert Hall with whom I share everything, including a love of history. His emotional and intellectual support as well as good humour has been essential to the completion of this project.

Carolyn Malone
December 2002

Abbreviations

General

BPDL	Barmaids' Political Defence League
CS	Certifying surgeons
FOLD	Freedom of Labour Defence Association
JCEB	Joint Committee on the Employment of Barmaids
LCWTOWRC	Lancashire and Cheshire Women's Textile and Other Workers' Representation Committee
MOH	Medical officer of health
NIPWSS	National Industrial and Professional Women's Suffrage Society
NUWSS	National Union of Women's Suffrage Societies
SPW	Society for Promoting the Employment of Women
TUC	Trades Union Congress
VA	Vigilance Association for the Defence of Personal Rights
WCG	Women's Cooperative Guild
WEU	Women's Emancipation Union
WIC	Women's Industrial Council
WIDC	Women's Industrial Defence Committee
WLF	Women's Liberal Federation
WLL	Women's Labour League
WTUL	Women's Trade Union League

Newspapers and journals

BMJ	*British Medical Journal*
CASW	*County Advertiser for Staffordshire and Worcestershire*
CC	*Common Cause*
CN	*Co-operative News*
DC	*Daily Chronicle*
ER	*Englishwoman's Review*
JSI	*Journal of the Sanitary Institute*
MG	*Manchester Guardian*
NS	*Northern Star*
PG	*Pottery Gazette*
PH	*Public Health*
PMG	*Pall Mall Gazette*
SS	*Staffordshire Sentinel*
WIN	*Women's Industrial News*

Archives

PRO	Public Record Office, Kew

Introduction

Between 1830 and 1914 a discourse of danger dominated public discussion of women's work outside the home. Over the course of time, different concerns and emphases came to the forefront of the debate; public discussions privileged different discourses at different moments. In each instance the definition of danger was based upon gender which I define, following Joan Scott, as socially constructed knowledge about sexual difference.[1] During the 1830s and 1840s, opponents of women's work highlighted its moral and sexual perils. The emphasis shifted in the 1870s to the neglect of children by working mothers and by the 1890s, the focus of this study, the debate turned on the dangers of physical degeneracy for women and death for their unborn children. The diverse public representations of the hazards of women's labour, in turn, shaped and influenced the creation of protective labour legislation in the 1840s, 1870s and 1890s.

During the 1890s the English government banned women from working in the most dangerous and highest-paying sections of the white lead trade. After considering a similar ban in the pottery trade, it mandated the medical inspection of women with suspension, without pay, if they exhibited signs of lead poisoning. The government could enact these policies because of a new avenue of protective labour legislation: dangerous trades regulation. The 1891 and 1895 Factory and Workshop Acts empowered the Home Secretary to declare a trade dangerous, enact special rules and limit or prohibit women's work within such trades.

Sensational stories in mass daily newspapers had drawn public and governmental attention to the perilous conditions that women and their unborn children faced in these two dangerous trades. Papers such as the *Daily Chronicle* horrified their readers with stories entitled 'White cemeteries: how women are poisoned', 'White cemeteries: massacre of the innocents', or 'Lead in the home: infanticide in the Potteries'. The *Star* and the other papers followed suit, graphically depicting the effect of lead on women's bodies, especially their reproductive organs, and recounting case after case of infant slaughter. They also published medical testimony that represented lead poisoning as a 'woman's problem', one that could be resolved only through the elimination of women from those harmful trades. These stories, which made their way into Home Office files, were the catalyst for extensive and in-

---

1 Joan Scott, *Gender and the politics of history*, New York 1988, 2.

tensive public debate and governmental investigations that culminated in the new and radical dangerous trades measures.

Since the emergence of gender as an analytical category in the field of labour history, there has been a growing historiography on protective labour legislation. For the most part, scholars have focused upon the extra-parliamentary campaigns of the 1840s and 1870s with particular emphasis upon the discourse on female labour that both prompted and legitimised special laws for them in the workplace.[2] They have also assessed the implications of those discourses and the resulting legislation for women workers. In contrast, there have been few works dealing with pre-war protection and they have primarily concentrated on the efforts to regulate sweated labour.[3]

Despite these important advances in scholarship, the most radical form of protection, dangerous trades regulation, remains to be explored. Focusing on the period of 1880 through 1914, this book analyses this crucial development in protective labour legislation and examines the discourse that centred upon the physical dangers of work for women and their unborn children. It is my contention that this was a new and more powerful way of conceptualising the social perils of women's work. Throughout the text I explore the critical links between language and the historical context because, following Ava Baron, I believe that language is not sealed off from the social world; rather social and political context shapes meaning.[4] The new journalism, medical writings on lead poisoning and the female body, the rise of labour, the expansion of feminist activism as well as imperial concerns for the future of the race were critical factors shaping both the public dialogue and resulting measures. Finally, I examine the role of government officials, politicians, the press, medical men, feminists and working men in this important episode in social policy-making. I am particularly concerned with why these historical actors supported or, in a few instances, contested the idea that certain work was especially dangerous for women. For, it was only after the discourse had the allegiance of individ-

---

[2] For an overview of the diverse interpretations see Robert Gray, *The factory question and industrial England, 1830–1860*, Cambridge 1996, 5–9. Important works on the subject include Sally Alexander, 'Women, class, and sexual difference in the 1830s and 1840s: some reflections on the writing of a feminist history', *History Workshop* xvii (1984), 125–49; Sonya O. Rose, *Limited livelihoods: gender and class in nineteenth century England*, Berkeley 1992; Deborah Valenze, *The first industrial woman*, New York 1995, ch. v; and Marianna Valverde, ' "Giving the female a domestic turn": the social, legal, and moral regulation of women's work in British cotton mills, 1820–1850', *Journal of Social History* iv (1988), 619–34.

[3] See Patricia Malcolmson, *English laundresses: a social history, 1850–1930*, Urbana 1986; Jenny Morris, *Women workers and the sweated trades: the origin of minimum wage legislation*, Aldershot 1986; and James A. Schmiechen, *Sweated industries and sweated labor: the London clothing trades, 1860–1914*, Urbana 1984. The exception to that statement is Barbara Harrison's work on gender and occupational illness, *Not only the 'dangerous trades': women's work and health in Britain, 1880–1914*, London 1996.

[4] Ava Baron, 'Gender and labor history: learning from the past, looking to the future', in Ava Baron (ed.), *Work engendered: toward a new history of American labor*, Ithaca 1991, 1–46.

uals, especially government officials, that its immense social and political implications were realised. In the end, I contend that the prevalent discourse of danger created a climate in which the government could, and did, enact protective measures that resemble what we today call 'foetal protection'.[5]

Chapter 1 briefly examines the origins and early issues of protective labour legislation. During the Short Time campaign of the 1830s and 1840s and the Nine-Hour Movement of the 1870s, parliamentary and extra-parliamentary proceedings focused upon women workers. In the 1830s and 1840s, as previously noted, opponents of women's work outside the home emphasised its moral and sexual perils. By the 1870s, however, they shifted their emphasis to the dangers that maternal neglect posed to young children. This chapter provides the critical background to the legislation of the pre-war period and explores how at different times different discourses were privileged in the public discussion of the dangers of women's work. It concludes with a very brief section on the 'discovery' of sweated labour in the late 1880s which led to renewed interest in protective labour legislation.

Following revelations about the extent of sweated labour, the House of Lords Select Committee on the Sweating System met between 1888 and 1890. Its investigation of the conditions of labour in one trade, the nail and chain trade, is the subject of chapter 2. Testimony that nail and chain work imperilled women's reproductive functions led the committee to recommend limitations on their work. Members of parliament for districts in which the trade was centred, as well as working men, then pressed the Home Secretary to create special measures restricting women's work in the trade. This chapter traces those developments and explores how they affected the woman worker. For, instead of creating special clauses to protect women in this one trade, the Home Secretary responded with a general clause in the 1891 Factory and Workshop Act that empowered him to declare a trade dangerous and, in consultation with the factory department, enact special regulations for it. This measure, intended to deal with the nail and chain trade, unexpectedly formed the basis of dangerous trades regulation.

Chapters 3 and 4 analyse the process of creating dangerous trades regulations for women in the white lead and pottery trades. They particularly highlight how the enactment of those regulations turned on interaction between the press, medical men, lobby groups and the government as well as ideas

---

[5] Two works on foetal protection in contemporary America led me to make this comparison: Cynthia R. Daniels, *At women's expense: state power and the politics of fetal rights*, Cambridge, Mass. 1993, and Barbara Duden, *Disembodying women: perspectives on pregnancy and the unborn*, Cambridge, Mass. 1993. Daniels has analysed the foetal protection policies enacted by Johnson Controls, Inc. during the 1980s that bear a striking resemblance to English policies enacted for the white lead trade in the 1890s. I have made this comparison in 'The gendering of dangerous trades: government regulation of women's work in the white lead trade in England, 1892–1898', *Journal of Women's History* viii (1996), 15–35.

about sexual difference and work. A series of articles in the *Daily Chronicle*, together with several government investigations, contributed to the development of a substantial campaign to remove women from the white lead trade. Convinced that he should take action, the Home Secretary proposed a clause in the 1895 Factory and Workshop Act that would empower him to limit or prohibit women's work in dangerous trades. This new power was subsequently exercised in the two lead trades under consideration. My analysis of the proceedings reveals several significant features of the construction of dangerous trades regulation. First, women workers protested against state intervention that would eliminate or limit their work opportunities in these high-paying jobs. Moreover, there was ample evidence that the conditions of the workplace were a critical factor in the development of lead poisoning as well as statistical evidence that these trades were hazardous to both women and men. In the end, these factors were overshadowed by an overriding concern for the protection of the potential offspring of these women workers.

Chapters 5 through 7 examine more extensively the ideas and activities of the press, medical men and feminists engaged in the process of creating dangerous trades regulations. Their emergence as the key actors in this process marks a significant moment in the history of protective labour legislation and warrants more in-depth analysis. Male workers and trade unionists had been the primary lobbyists for restrictions on women's labour during the 1840s and 1870s. As chapter 2 will illustrate, they continued to be very vocal proponents of measures to restrict female labour in the nail and chain trade in 1891. However, their participation was drastically reduced as state intervention moved in the direction of constructing measures to safeguard women's reproductive health. Male white lead workers were absent from the discussion of women's work in their trade, while there were relatively few comments on this subject made by male pottery workers. Resolutions supporting the extension of the Home Secretary's power to restrict women's work in dangerous trades were passed at the Trades Union Congresses in 1894 and 1895 while lead poisoning was later discussed in connection with the extension of workmen's compensation to occupational illnesses. During this period, it may be argued, dangerous trades regulations figured only minimally on this organisation's agenda, an agenda that was centrally concerned with issues such as the eight-hour day or workmen's compensation.[6] Instead, the Women's Trade Union League took the lead and diligently pursued state protection for women working in dangerous trades.

---

[6] Peter Bartrip has argued that during the 1880s and 1890s the Trades Union Congress was responsible for drawing up a 'rash' of employer liability bills that came before parliament: 'The rise and decline of workmen's compensation', in Paul Wiendling (ed.), *The social history of occupational health*, London 1985, 157–79 at pp. 160–1. See also Peter Bartip and S. B. Burman, *The wounded soldiers of industry: industrial compensation policies, 1833–1897*, Oxford 1983, 158–206.

INTRODUCTION

In contrast, the press undertook a series of investigations to uncover and publicise the hazards of women's work in the lead trades. And, it should be noted, its coverage of female labour extended far beyond the lead trades to include the nail and chain, match and mining trades. The proliferation of such stories is an important but neglected facet of the 'new journalism' of the late 1880s.[7] Sensational stories about women's work provide a way of exploring new perspectives on the rise of the mass, daily newspaper. First, editors and journalists believed that these stories resonated with their readers and sold newspapers. Second, they illustrate that papers were trying to carry out William T. Stead's idea of 'government by journalism'. He claimed that the press could and should wield its immense power to uncover social problems and promote reform. Finally, I believe that these stories are a variation of what, in *City of dreadful delight*, Judith Walkowitz has called narratives of sexual danger.[8] They contributed to and reinforced the idea that women who crossed prescribed social boundaries placed themselves at risk; in this case, they faced bodily danger. In the end, newspapers played a decisive role in the contest over women's position in the workplace and in public as well as in the establishment of legal boundaries for them in those spaces.

Doctors were, according to Frank Mort, prominent in the growing state scrutiny of women and in the formation of social policies affecting them in the years prior to the First World War.[9] Peter Bartrip has demonstrated that this was certainly the case regarding the creation of dangerous trades regulations.[10] They acquired a prominent place in the factory department with the appointment of Dr Benjamin Arthur Whitelegge as the chief inspector of factories in 1896 and Dr Thomas Legge as the first medical inspector of factories two years later. The responsibilities of certifying surgeons, who had been examining and certifying that children seeking factory employment were fit, expanded dramatically during the 1890s. As a result of the 1895 Factory and Workshop Act, they examined workers suspected of suffering from lead poisoning and they reported confirmed cases to the government. They also conducted the monthly medical examination of women working the pottery trade instituted by the government in 1898. The routine use of their reports

---

[7] The most valuable work on the subject is Joel H. Weiner (ed.), *Papers for the millions: the new journalism in Britain, 1850s to 1914*, Westport 1988. See also Lucy Brown, *Victorian news and newspapers*, Oxford 1985; Stephen Koss, *The rise and fall of the political press in Britain: the nineteenth century*, Chapel Hill 1981; and Alan J. Lee, *The origins of the popular press, 1855–1914*, London 1976.

[8] Judith Walkowitz, *City of dreadful delight: narratives of sexual danger in late Victorian England*, Chicago 1992.

[9] Frank Mort, *Dangerous sexualities: medico-moral politics in England since 1830*, New York 1987.

[10] Peter Bartrip, 'Expertise and the dangerous trades, 1875–1900', in Roy MacLeod (ed.), *Government and expertise: specialists, administrators and professionals, 1860–1919*, Cambridge 1988, 89–109.

by Legge further expanded their influence.[11] Finally, medical officers of health, whose appointment had been mandated by the 1872 Public Health Act, also collected and contributed valuable information about industrial diseases.[12] The state relied extensively upon the expertise of this corps of medical men when it decided to enact dangerous trades regulations. And, very significantly, they were at the forefront of the construction of the idea that certain work was especially harmful for women and their unborn children. This point was publicised in the press and repeatedly developed in their writings, official reports for the government and testimony before governmental committees. In chapter 6 I give the writings of doctors on women and lead poisoning a close and critical reading; I argue that their emphasis on sexual difference as the key determinant of the disease was a new and questionable theory. For, despite statistical evidence that unsafe working conditions made men equally liable to illness and death, Dr Thomas Oliver and others maintained that lead poisoning was a 'woman's problem'. I contend that their conceptualisation of the problem was shaped by the fact that they were sexual scientists who believed that nature outfitted women to be mothers, not workers. Concerned also with the future of the race, Oliver expressed the sentiments of many of his colleagues when he characterised lead as a racial poison. Their writings substantiate the argument of Thomas Laqueur and other historians about the gendered nature of science.[13]

The creation of dangerous trades regulations was a divisive, controversial issue for the feminist organisations that are the subject of chapter 7. This book substantiates and amplifies Rosemary Feurer's point that the passage of sex-specific labour legislation was a source of contention between the so-called 'social feminists' in groups such as the Women's Trade Union League and Women's Labour League and 'equal-rights feminists' belonging to the Society for Promoting the Employment of Women and the Freedom of Labour Defence Association.[14] They were at opposite ends of the debate over

---

[11] Jeanne L. Brand, *Doctors and the state: the British medical profession and government action in public health, 1870–1912*, Baltimore 1965, 145.

[12] Ibid. 126–9.

[13] See, for example, Thomas Laqueur, *Making sex: body and gender from the Greeks to Freud*, Cambridge, Mass. 1990; Brian Easlea, *Science and sexual oppression*, London 1981; Elizabeth Fee, 'Science and the woman problem in historical perspective', in M. S. Teitelbaum (ed.), *Sex differences: social and biological perspectives*, Garden City, New York 1976; Jill Conway, 'Stereotypes of femininity in a theory of sexual evolution', in Martha Vicinus (ed.), *Suffer and be still*, Bloomington 1973, 140–54; Ludmilla Jordanova, *Sexual visions: images of gender in science and medicine between the eighteenth and twentieth centuries*, Madison 1989; Susan Sleeth Mosedale, 'Science corrupted: Victorian biologists consider "the woman question"', *Journal of the History of Biology* xi (1978), 1–56; Cynthia Eagle Russett, *Sexual science: the Victorian construction of womanhood*, Cambridge, Mass. 1989; Londa Schiebinger, *The mind has no sex? Women in the origins of modern science*, Cambridge, Mass. 1991; and Nancy Tuana, *The less noble sex: scientific, religious, and philosophical conceptions of women's nature*, Bloomington 1993.

[14] I have adopted the terminology used in Rosemary Feurer, 'The meaning of "sisterhood":

the regulation of women's work in the lead trades as well as further campaigns to prohibit women's allegedly dangerous work as barmaids (1906 and 1908) and pit-brow workers at coal mines (1911). This chapter examines the latter episodes as case studies of feminists' divergent perspectives on protective labour legislation, perspectives shaped by political orientation as well as beliefs about the supremacy of gender or class solidarity. Very significantly, feminists who contested the proposed bans on barmaids and pit-brow workers undertook their own investigations to prove that such labour would not harm women or their unborn children. This strategy proved to be successful in blocking the proposed measures and reveals that the label 'dangerous trade' was frequently applied to jobs that were not inherently dangerous for women but were considered unsuitable for them.

In the epilogue I place the movement to protect women and their unborn children in the national and international context. I suggest that dangerous trades regulation was an important, but historically neglected, facet of a larger pre-war campaign to safeguard the 'future of the race'. As England became increasingly caught up in economic and imperial rivalry, especially after 1900, the quality and quantity of its population became an urgent political issue. As numerous historians have shown, many government officials viewed maternal ignorance and negligence as the cause of the population problem; their concerns, in turn, led to a proliferation of maternal and child welfare schemes.[15] It is my contention that legislative attempts to limit or remove women from dangerous work was another way in which the state tried to address the population problem. I will also briefly compare my analysis of protection in pre-war England with developments in other European countries. In pre-war France and Germany, a similar recasting of the woman worker problem took place as the legislatures in those countries enacted laws restricting women's work opportunities in order to serve the national

---

the British women's movement and protective labor legislation, 1870–1900', *Victorian Studies* xxxi (1988), 233–60; Malcolmson, *English laundresses*; and Ellen Mappen, 'Strategies for change: social feminist approaches to the problems of women's work', in Angela V. John (ed.), *Unequal opportunities: women's employment in England, 1800–1918*, Oxford 1986, 235–59.

[15] See Anna Davin, 'Imperialism and motherhood', *History Workshop* v (1978), 6–66; Deborah Dwork, *War is good for babies and other young children: a history of the infant and child welfare movement in England, 1898–1918*, London 1987; Jane Jenson, 'Gender and reproduction: or, babies and the state', *Studies in Political Economy* xx (1986), 9–46; Jane Lewis, *The politics of motherhood: child and maternal welfare schemes in England, 1900–1939*, London 1980, and 'The working-class wife and mother and state intervention, 1870–1918', in Jane Lewis (ed.), *Labour and love: women's experience of home and family, 1850–1914*, Oxford 1986, 99–120; Sonya Michel and Seth Koven, 'Womanly duties: maternalist politics and the origins of welfare states in France, Germany, Great Britain, and the United States, 1880–1920', *American Historical Review* lxxxxv (1990), 1076–108; and Ellen Ross, *Love and toil: motherhood in outcast London, 1870–1918*, Oxford 1993.

interest.[16] This was a problematic resolution for the objects of this legislation, for, as governments viewed women solely as potential mothers, they were deprived of the opportunity to make decisions about their work and reproduction.

[16] Several excellent essays on this subject are found in Elinor A. Accampo, Rachel G. Fuchs and Mary Lynn Stewart (eds), *Gender and the politics of social reform in France, 1870–1914*, Baltimore 1995, and Laura L. Frader and Sonya O. Rose (eds), *Gender and class in modern Europe*, Ithaca 1996. See also Kathleen Canning, *Languages of labor and gender: female factory work in Germany, 1850–1914*, Ithaca 1996, and Mary Lynn Stewart, *Women, work, and the French state: labour protection and social patriarchy, 1879–1919*, Montreal 1989.

1

'Discovering' the Dangers of Women's Work, 1830–1880s

The crowding together of numbers of the young in both sexes in factories, is a prolific source of moral delinquency. The stimulus of the heated atmosphere, the contact of the opposite sexes, the example of the lasciviousness upon the animal passion – all have conspired to produce a very early development of sexual appetencies: Peter Gaskell, *The manufacturing population of England* (1833)[1]

The prolonged absence from home of the wife and mother caused an enormous amount of infant mortality and it must cause the elder children to be more or less neglected. It deadened the sense of parental responsibility: Thomas Maudsley, secretary, Committee for Promoting the Nine Hours Movement (1872)[2]

From a purely physical point of view the nation's strength is measured by its reproductive power and the high percentage of the fitness of its children. . . . Women's work becomes the cause of physical degeneracy and of inability on the part of women to rise to the dignity of the completed act of motherhood: Dr Thomas Oliver, Lecture before the Eugenics Education Society (1911)[3]

Each of these statements was made as part of the public debate about enacting protective labour legislation in England. They were diverse manifestations of a single idea, the idea that women's work outside the home was dangerous to society and required state intervention. Between 1830 and 1914 a discourse of danger dominated the public discussion of female labour yet, as the opening quotations suggest, different types of danger were emphasised at various times. In the 1830s and 1840s the emphasis was upon the sexual hazards of their work outside the home. From the 1870s through 1914, in contrast, women's labour was considered dangerous primarily because it prevented them from fulfilling their maternal responsibilities; impeding either their ability to care for their young children or their ability to produce children at all.

---

1 Peter Gaskell, *The manufacturing population of England*, London 1833; repr. New York 1972, 68.
2 Proceedings of a meeting at the Manchester Town Hall, 20 July 1872, reprinted in the MG, 22 July 1872.
3 Dr Thomas Oliver, 'Lead poisoning and the race', *BMJ*, 13 May 1911, 1096.

'Discourses', Ava Baron has written, 'compete with each other for the allegiance of human agents. The agency of individuals is required before the social and political implications of a discourse can be realized.'[4] As the various discourses on the dangers of women's work gained the allegiance of individuals, their social and political implications for women were realised through the creation of special labour laws. Initially limiting the hours of women's work in factories, the locales of protection expanded throughout the century to encompass workshops, home work and, ultimately, women's bodies. The 1847 Factory Act limited women's work in textile factories to ten hours per day while the working day was shortened to nine hours by the 1874 Factory Act. The 1891 and 1895 Factory and Workshop Acts included provisions to combat sweated labour in workshops and homes, prohibited women from working four weeks after childbirth and, most significantly, created the apparatus for regulating their labour in trades considered especially dangerous for them. The government was empowered to declare a trade dangerous, enact special rules, and limit or prohibit women's work within them. It exercised that unprecedented power and enacted policies which, I suggest, resemble contemporary 'foetal protection' measures.

This chapter examines the public discourses on the dangers of women's work during the period from 1830 through the 1880s and their critical link to the creation of protective measures for them. The particular representations of the hazards of women's work were based upon gender, that is, knowledge about sexual difference. This knowledge, Joan Scott and Mary Poovey have persuasively argued, was not absolute or true but socially constructed and open to revision and opposition.[5] Thus, I argue, it is essential to analyse the specific historical contexts to understand how and why different aspects of this knowledge were prominent in the languages of protection.[6] The transformations within the discourse of danger, from an emphasis on sexual dangers to maternal dangers, were due to specific economic, social, political and intellectual developments. The very public presence of female labour in factories, the rise of the medical profession and feminist organisations, organised labour activities, Malthusian ideology and scientific writings on sexual difference were seminal in this respect.

---

[4] Baron, 'Gender and labor history', 31.
[5] Scott, *Gender and the politics of history*; Mary Poovey, *Uneven developments: the ideological work of gender in mid-Victorian England*, Chicago 1988.
[6] Critics of poststructuralist language analysis have rightly argued that the exclusive emphasis on discourse has produced ahistorical studies devoid of human agency. However, several historians have made significant points that serve as correctives to this problem and inform my approach. Most significantly, they have emphasised the essential relationship between text and context; between historical actors, their experiences and human agency. For an excellent overview of the impact and critique of this approach see Baron, 'Gender and labor history', and Kathleen Canning, 'Feminist history after the linguistic turn: historicizing discourse and experience', *Signs* xix (1994), 368–404.

This chapter's analysis provides the critical background upon which the legislation of the 1890s through 1914 was built. Most significantly, it enables me to illustrate that the pre-war emphasis on work and reproductive danger, the major contention of this book, represented another, and ultimately more powerful, shift in the conceptualisation of the hazards of women's work.

Sex in industry

During the 1830s and 1840s a coalition of philanthropists, mill owners, Tory radicals and working men formed short-time committees and agitated for the limitation of the legal working day. The crowning achievement of this movement was the 1847 Factory Act which officially limited women's and children's (and unofficially men's) work in textile factories to ten hours per day.[7] Participants critiqued the factory system for a variety of reasons, including its effect on health and reproduction, wage levels, and the hazards of the new machinery.[8] However, as several historians have shown, the discourse on women's labour particularly centred upon its sexual implications.[9]

Contemporary documents reflect the belief that the mixing of the sexes in factories and mills led to pre-marital or extra-marital sex, illegitimacy and sexual exploitation.[10] Middle-class observers, including Peter Gaskell quoted at the beginning of this chapter, harped on these disastrous results. He claimed that the heated atmosphere, together with close contact between the sexes, promoted early sexual activity. He labelled factories 'hotbeds of lust', places of 'unbridled indulgences' where chastity was a 'laughingstock'.[11] Friedrich Engels criticised factory girls after he collected statistics showing that three-quarters of them between the ages of fourteen and twenty were unchaste.[12] Working men also put forward a critique of the intermixing of men and women in the factories. John Doherty, leader of the Manchester short-time committee and of an all male mule-spinners union, said that 'if he [a Glasgow manufacturer] could not find in his heart to employ, and pay men for doing his work, he should look out for women whose morals are already corrupted, instead of those whose lives are yet pure and spotless'.[13] James

---

7 The movement for the ten-hour day has generated tremendous interest among historians who have interpreted it in several different ways. For an overview of the debate see Gray, *The factory question*, 5–9.
8 Ibid. 1–159.
9 See Alexander, 'Women, class, and sexual difference', 125–49, and Valverde, ' "Giving the female a domestic turn" ', 619–34.
10 Jane Humphries, ' "... The most free form of objection ...": the sexual division of labour and women's work in nineteenth-century England', *Journal of Economic History* xlvii (1987), 929–48.
11 Gaskell, *The manufacturing population*, 63–5.
12 This point was made in Valenze, *The first industrial woman*, 99.
13 Alexander, 'Women, class, and sexual difference', 137.

McNish, a prominent leader of the spinners' union in Scotland, said that he knew only a few girls from his department who were not prostitutes. It was his opinion that bringing young women together in industrial establishments tended 'to render them vicious and dissolute and to demoralize them'.[14] Other working men represented working women as the victims of sexual predation by management. Thus, as a result of the extensive employment of women in factories, powerloom weaver Richard Pilling complained that 'the overlookers, managers, and other tools, take the most scandalous liberties with them'.[15] John Deegan, a card room hand who opposed women's factory labour, also spoke of them as being 'polluted by lick-spittles who are placed over them in factories and coal-pits'.[16]

Why did men across class lines employ a sexual discourse in their critique of female employment in factories and mills? 'Everything in the new social order', Thomas Laqueur has argued, 'was heated up, changeable, morally shaky and sex was the prism through which its dangers were imagined.'[17] As a result, a sexual discourse was present in a variety of rhetorical arenas in the 1830s and 1840s. Upper- and middle-class commentators were responding to the emergence of a new social order in which older restraints no longer operated to control the working class. It was mobile, sexually active and multiplying in numbers. Such apprehensions about sexuality and population growth were influenced, above all, by the writings of Thomas Malthus.[18]

Malthus' assertion that 'the passion between the sexes is necessary and will remain nearly in its present state', combined with the belief that the lower orders lacked the self-discipline to refrain from sexual activity, strongly impressed his contemporaries. The image of an unruly and highly sexualised working class was reflected in the reports of social reformers and utilitarians, such as James Kay and Edwin Chadwick. Referring to the working class, Kay wrote 'There is . . . a licentiousness capable of corrupting the whole body of society. . . . Sensuality has no record.'[19] Thus, sexual analysis was pervasive in

---

14 Valverde, ' "Giving the female a domestic turn" ', 629.
15 *The trials of Feargus O'Connor and fifty-eight others charged with sedition*, Manchester 1843; repr. New York 1970, 253.
16 John Deegan, a Chartist, spoke at a public meeting at Leicester and the proceedings were reported in the *NS*, 1 June 1839.
17 Thomas Laqueur, 'Sex and desire in the industrial revolution', in Patrick O'Brien and Roland Quinault (eds), *The industrial revolution and British society*, New York 1993, 100–23 at p. 118.
18 See, for example, Catherine Gallagher, 'The body versus the social body in the works of Thomas Malthus and Henry Mayhew', in Catherine Gallagher and Thomas Laqueur (eds), *The making of the modern body: sexuality and science in the nineteenth century*, Berkeley 1987, 83–106; Laqueur, 'Sex and desire'; and Valenze, *The first industrial woman*, 128–80.
19 Dr James Kay, *The moral and physical conditions of the working classes in the cotton manufacture in Manchester*, London 1832; repr. London 1970, 62.

public health investigations, poor law reports and commentary on sanitation and overcrowding as well as in the factory debate.[20]

In the new industrial world, Sally Alexander has argued, working-class men felt 'threatened in their whole being with loss of skill, sexual, and economic authority'.[21] Changes in the organisation of production, Jane Humphries has further emphasised, entailed the movement of women and girls from domestic industry to mills and factories. This meant a loss of family, or rather paternal, authority and control over their work and sexuality.[22] A classic example of this transformation was the case of the transition from handloom weaving in the cottage to powerloom weaving in the factory. A Stockport powerloom weaver, Thomas Leonard, suggested that women's 'inefficiency' at such work placed them at the mercy of the lustful overlookers. As a result, he claimed, many young women were reduced to public prostitution. If, however, more men worked in the weaving shed they could protect innocent young females from sexual predation and restore sexual order.[23]

A view of working-class women as passionate beings with a strong sexual drive was implicit in these representations of sexual danger. It was also explicitly stated by some contemporaries like Richard Carlile who said 'No one shall persuade me but that healthy girls, after they pass through the period of puberty, have an almost constant desire for copulation.'[24] This natural disposition, in turn, made them more susceptible to seduction and wanton behaviour. This depiction of women's nature, of course, had dominated western thought for centuries but now it carried with it class connotations borne of the emergence of separate sphere ideology.[25] This increasingly dominant ideology characterised middle-class women as 'angels in the home', eminently moral and distinctly asexual beings.[26]

---

[20] For an analysis of the anxiety of reformers Kay and Chadwick see Poovey, *Making a social body*, 55–72, 98–131.
[21] Alexander, 'Women, class, and sexual difference', 137.
[22] Humphries, ' "... The most free form of objection." ', 930, 943–4. This was also evident in the rhetoric of the Chartist campaign. See also Anna Clark, 'The rhetoric of Chartist domesticity: gender, language, and class in the 1830s and 1840s', *Journal of British Studies* xxxi (1991), 62–88, and Robert G. Hall, 'Unsexing the male: gender, technology, the state, and Chartism in the cotton district, 1830–1860', paper presented at the 1993 annual meeting of the Southern Conference on British Studies.
[23] Meeting of powerloom weavers in Stockport, reprinted in the NS, 2 May 1840.
[24] Richard Carlile to Francis Place, 8 Aug. 1822, BL, Francis Place newspaper collection, vol. 68. For the place of Carlile's sexual ideology within plebeian culture see Anna Clark, *The struggle for the breeches: gender and the making of the British working class*, Berkeley 1995, 179–85.
[25] See Laqueur, *Making sex*, ch. v, and Tuana, *The less noble sex*, ch. vii.
[26] This important development has been analysed in Nancy F. Cott, 'Passionlessness: an interpretation of Victorian sexual ideology, 1790–1850', *Signs* iv (1978), 219–36; Catherine Hall, *White, male, and middle class: explorations in feminism and history*, New York 1992,

These ideological orientations contributed to bringing concerns about sexual immorality to the forefront of the discussion of the dangers of women's work in the 1830s and 1840s. They were also the basis for arguments for the prohibition or legal restriction of women's work in mills and factories. That course of action, it was argued, would eliminate sexual danger in the workplace and restore what was repeatedly referred to by male participants as the natural order.[27] In the end, parliament decided not to exclude women but rather to limit their working day. This was an important starting-point from which later discussions of the woman worker problem and protective labour legislation would evolve.

Maternal neglect

The success of the short-time campaign and the collapse of the Chartist movement ushered in a politically calmer atmosphere, the mid-century equipoise. The waning of this period of relative social harmony was, however, followed by a growth in union activity and localised initiatives among working men to improve the conditions of their labour. Beginning in the late 1850s, a variety of unions successfully and unsuccessfully struck for the nine-hour day. In 1872 the Factory Acts Reform Association was formed for the purpose of reducing hours in the cotton trade. It attracted the support of A. J. Mundella, MP for Sheffield, who introduced bills in 1872 and 1873. In the face of mounting pressure, the government investigated conditions of work in cotton factories and introduced its own bill which subsequently became law in 1874. The 1874 Factory Act reduced the working day in textile factories to nine hours. The law was drafted in terms of women and children's hours but, as in 1847, it had the effect of shortening hours for men as well.

The process of creating this legislation has been well analysed by Sonya Rose.[28] She has demonstrated that men made no mention of seeking shorter hours for women and children before they sought parliamentary assistance. It was only when they failed in private negotiations with their employers and went 'public' that they spoke on their behalf. This, of course, was an astute tactical shift since the state was relatively unwilling to intervene in men's work.[29] Most significantly for this chapter, Rose has deconstructed the public

---

75–107; Susan Kingsley Kent, *Sex and suffrage in Britain, 1860–1914*, Princeton 1987, 24–59, 80–113; and Poovey, *Uneven developments*, 1–23.

[27] The complaints about women's work violating the natural order have been discussed in Valverde, ' "Giving the female a domestic turn" ', 627–8, and Valenze, *The first industrial woman*, 100–1.

[28] Rose, *Limited livelihoods*, ch. iii.

[29] This point had previously been made by Rose in ' "From behind the women's petticoats": the movement for a legislated nine hour day and state protection of working women in

debate and persuasively argued that it and the resulting law 'were cultural productions which contributed to the formulation of the social problem of "the working mother" '.[30]

This new emphasis on the relationship between work and maternity was conspicuously evident in the government's approach to the factory question in the 1870s. Dr Bridges and Dr Holmes made perhaps the most significant contribution to this development when their government-commissioned report was published in 1873. Explicitly told to focus on the effect of married women's work on their offspring, they asked a large group of medical men if it increased the rate of infant mortality. Out of the 132 doctors they surveyed, 101 said that it did. One factory surgeon set forth what Bridges and Holmes considered the general opinion: 'I regard the mother's return to the mill as almost a death sentence to the child.'[31] Another medical man responded, 'I believe if married women were kept at home to attend their houses, nine-tenths of the evils in the factory districts would be removed.'[32] Bridges and Holmes concluded that 'Those of our correspondents who see no evidence of increased mortality reside chiefly in country districts or in parts of the country where suckling women do not generally work.'[33] In view of this evidence, they recommended the reduction of women's and children's hours to fifty-four per week. As for married women (whom they assumed were mothers), they endorsed some arrangement 'by which mothers of young infants shall either be employed half time or be excluded for a time from the factories altogether'.[34] A variety of proposals for the restriction of married women's work, including their reduction to half-time work or mandatory maternity leaves, were indeed subsequently pursued in parliament.[35]

When working men pressed their cause they, too, employed the terms of the current woman worker debate and played upon prevailing social concerns. In July 1872 Thomas Maudsley, secretary of the Committee for Promoting the Nine Hours Movement, sent a paper to the Home Office entitled, 'Labour in the cotton mills, and the necessity for shorter hours for women and young people'. Women working in the cotton factories, he

---

Britain, 1870–1878', *Journal of Historical Sociology* ix (1991), 32–51, and 'Gender antagonism and class conflict: strategies of male trade unionists in nineteenth-century Britain', *Social History* xii (1988), 191–208.
30 Idem, *Limited livelihoods*, 59.
31 J. H. Bridges and T. Holmes, *Report to the Local Government Board on proposed changes in the hours and ages of employment in textile factories*, PP 1873, [c.754] lv. 840.
32 Ibid.
33 Ibid. 841–2.
34 Ibid. 863.
35 The various proposals were discussed and criticised in Whately Cooke-Taylor, 'What influence has the employment of mothers in manufactures on infant mortality; and ought any, and what, restrictions to be placed on such employment?', in Charles Wager Ryalliss (ed.), *Transactions of the National Association for the Promotion of Social Sciences*, London 1874, 569–85 at pp. 573–4. See also Rose, *Limited livelihoods*, 64–7.

argued, 'have not the same privilege of attending to their duties as most other female operatives, thus often causing their infants to be sorely neglected. If a mother overlays herself, for fear of being discharged or abated, she rushes off to her work without attending to the wants of her infant'.[36] That same month he made the charges quoted at the beginning of this chapter at a meeting in Manchester in support of the nine-hour day. In a letter to parliament, the Factory Acts Reform Association responded to the charge that they were acting out of self-interest by claiming that 'Where women are employed, there is always a risk that infant life will suffer; and if this comes to pass, the health of the coming generation is exposed to danger.'[37] According to the working men, shortening the working day would eradicate the problem of infant mortality.

In striking contrast to what had happened in the 1840s, middle-class women's rights groups now mounted an organised campaign against the enactment of further special legislation for women. The SPW, led by Jessie Boucherett, and the Vigilance Association for the Defence of Personal Rights, led by Josephine Butler, were the major participants in the heated campaign against protection. The former group, founded in 1859 in order to provide education and training for middle-class women seeking work, also lobbied to prevent the passage of further labour laws, which they viewed as encroachments on working women's work opportunities.[38] The latter group, formed in 1871, was the most important offshoot of the National Association and the Ladies National Association, which had campaigned for the repeal of the Contagious Diseases Acts.[39] It sought to repeal or amend all laws that violated 'the principle of the perfect equality of all persons before the law, irrespective of sex or class'.[40]

Operating within the framework of orthodox liberal ideology, each group called for the application of its principles to women. Proposals for mandatory maternity leaves particularly elicited charges that women's rights were being violated by the state. The VA argued that such enactments exceeded parliament's power. 'It lies very near to the domain of personal right', the group claimed, 'including the limited sovereignty we possess over our own person, into which laws can seldom interfere or prudence without injustice.'[41] Moreover, the legislators and the public were concerning themselves with the

[36] Thomas Maudsley (letter and paper), to the Home Office, 29 Apr. 1872, PRO, HO 45/9308/12500.
[37] MG, 11 June 1873.
[38] The SPW's aims were outlined in 'Statement of the Association for the Employment of Women', *English Woman's Journal* iv (1859), 54–9.
[39] For more on the activities of the Ladies National Association in respect of the Contagious Diseases' Campaign see Judith Walkowitz, *Prostitution and Victorian society: women, class, and the state*, Cambridge 1980.
[40] VA, *Constitution and rules*, Manchester 1871.
[41] The point was made by the VA in a letter to the editor of the *Manchester Examiner and Times*, 20 June 1874.

welfare of the children while ignoring the other person involved – the mother. 'We do not think it conducive to the growth of wholesome family relationships', they maintained, 'to act upon the maxim that a woman is merely a piece of child-bearing mechanism, and that all faith in her affection and all regard for her rights are to be set aside out of consideration of her offspring.'[42] While not disputing the problem of infant mortality, they refused to blame the women and offered different interpretations of the problem and the likely consequences of the proposed solutions.[43] Boucherett repeatedly claimed that this legislation, by eliminating women from job competition, would only serve the interests of working men.[44] Even worse, she and Butler claimed that women would resort to prostitution to earn a livelihood.[45]

Several factors, I argue, converged to create the new maternal-centred discourse. Public interest had turned to the protection of infants in the late 1860s with the extensively publicised murder trials of Margaret Waters and Sarah Ellis. Charged with the deaths of several infants left in their care, their cases illustrated the perils of 'baby-farming'.[46] Concern over infanticide and infant mortality was further fuelled by the reports by local MOHs in manufacturing districts and articles in medical journals such as *The Lancet*. These were filled with comments like those made by Dr Greenhow who said that children of working women were left to nurses and

> superadded to the loss of their natural food during the greater part of the day, these poor babies are deprived of the warmth and comfort of their mothers' bosom, and it may likewise be added of the active exercise in which healthy children delight, and which is so conducive to their health and to the proper development of their muscular system.[47]

He, like most of the other MOHs, blamed excessive rates of infant mortality on women's work.[48] The 1873 registrar-general's report on births, deaths and

---

[42] Ibid.
[43] See, for example, *Proposed legislative restrictions upon the labour of women*, London 1874, *The factory (health of women) bill*, London 1874, and *The right of women to labour*, London 1874.
[44] See, for example, Jessie Boucherett, 'Legislative restriction on women's labour', ER iv (1873), 252–8.
[45] See idem, 'Events of the quarter – Mundella's bill and shop hours regulation bill', ER iv (1873), 209–11, and Josephine Butler, 'Letter to the editor', MG, 5 May 1874.
[46] Public concern was manifested, for example, in the formation of pressure groups such as Dr J. B. Curgeven's Infant Life Protection Society and the government's appointment of a Select Committee on the Best Means of Protecting Infants put out to Nurse in 1871. The following year the first Infant Life Protection Act was passed requiring the registration of births and deaths of infants in baby-farming institutions.
[47] *Report of the medical officer of the privy council for 1861*, PP 1862, [c.467] xxii. 655.
[48] Margaret Hewitt has presented a full listing of papers written on infant mortality in Lancashire in her study *Victorian wives and mothers*, London 1958. She has argued that the

marriages in Britain contributed further evidence of the dangers of women's work. Dr Farr, of the registrar's office, made a list of now familiar indictments against working mothers: their children were rarely nursed, fed artificial food, given cordials and often deserted. He attributed the high rate of women deserting their children and families to the high wages they earned in mills.[49] Given the focus of these important reports, it is not surprising that the 'working mother' was the centre of the public discourse.

This recasting of the working-woman problem would also seem to correspond to a shift in perspectives on their nature. It is very significant that charges of sexual misconduct were much less conspicuous than previously. This relative silence on the part of both middle- and working-class men represents a dramatic departure from the 1830s and 1840s. It signals the extent to which the idea of the 'passionless' female had become firmly implanted in the popular and prescriptive literature of the 1860s and 1870s. Dr William Acton's perspective on the ideal English wife is representative of contemporary views.[50] 'I am ready to maintain', he wrote in 1875,

> that there are many females who never feel any sexual excitement whatever. Others, again, immediately after each period, do become to a limited degree, capable of experiencing it; but this capacity is temporary. . . . Most of the best mothers, wives, and managers of households know little of or are careless about sexual indulgences. Love of home, of children, and domestic duties are the only passions they feel.[51]

The debate over women's work in the 1870s provides evidence that this respectable definition of womanhood was being extended to working-class women as well. As a result, their previous definition as passionate and sexual beings was being transmuted and displaced in the discourse on factory labour.

Within this context working women were problematic because they did not fulfil society's expectations of proper motherhood. Employed outside the home, they departed from social norms and this departure elicited tremendous commentary. Their work was perceived to be the cause of the egregious social problem of infant mortality. Proposals to eliminate or severely restrict

---

desire to remove women from the mills led observers to present a conflated picture of the problem.
[49] *Thirty fourth annual report of the registrar-general of births, deaths, and marriages in England*, appendix: 'Letter to the registrar-general on the cause of death in England in 1871', by William Farr, Esq, MD, FRS, PP 1873, [c. 806] xx. 225–9.
[50] This point, widely accepted by historians of gender and medicine, has been disputed in Michael Mason, *The making of Victorian sexuality*, New York 1994, 195–205. He argues that this remark is without parallel in the medical literature and inconsistent with Acton's views of female sexuality as expressed in other works.
[51] An excerpt from Dr William Acton, *The functions and disorders of reproductive organs, in childhood, adult age, and advanced life, considered in their physiological, social, and moral relations*, has been reprinted in Janet Horowitz Murray (ed.), *Strong minded women and other lost voices from nineteenth century England*, New York 1982, 127–9 at p. 127.

married women's labour were defeated because of the persistence of *laissez-faire* attitudes among members of parliament; however, the public discussion raised the spectre of possible regulation of women's reproductive as well as productive lives.

The sweating crisis of the 1880s

After a brief lull, public and governmental interest in the regulation of women's work re-emerged in the late 1880s, within the context of the newly uncovered sweating crisis.[52] Characterised by long hours of work, low pay and bad working conditions, sweating was the subject of sensational articles such as Arnold White's 'The nomad poor of London', published in 1886.[53] Unions and labour organisations pressed for the amendment of the existing labour laws to include regulation of domestic workshops and home work, which appeared to be the sites of such exploitation. The TUC passed resolutions calling for the amendment of the existing factory legislation to combat the problem in 1887 and 1888.[54] Questions in the House of Commons and in the Lords increased pressure on the government to act.[55] As a result, the government commissioned an investigation by members of the factory department and, most important, established the House of Lords Select Committee on the Sweating System in 1888.[56] This committee investigated the conditions of labour in twenty-seven sweated trades in London and the provinces over the course of the next two years. As the next chapter will illustrate, revelations of sweated labour in the nail and chain trade would unexpectedly lead to a further transformation within the long-standing discussion of the dangers of women's work. In this case, the debate would turn upon the question of the impact of sweated labour upon women's bodies or, to be more precise, upon their reproductive organs.

---

[52] For more on sweating see Morris, *Women workers*, and Schmiechen, *Sweated industries and sweated labor*.
[53] This point was made in Morris, *Women workers*, 7.
[54] *Annual report of the TUC* (1887), 43; (1888), 39–40.
[55] Morris, *Women workers*, 8.
[56] Home Secretary Matthews ordered the factory department to investigate the workshops under its jurisdiction in the East End of London in December 1887. The chief inspector of factories, Robert Redgrave, completed this report on 18 Feb. 1888: HO 45/9773/B1508.

2

Nails, Chains and Reproduction

Women using an oliver, a hammer operated by a treadle to make nails and chains, a witness told the House of Lords Select Committee on the Sweating System were very 'liable to misplacement of the womb, and to rupture, and also among the married women, I find they are very liable to miscarriages, as they frequently go on working when they are in the family way'.[1] This was one of several complaints made about female labour in the trade and, not surprisingly, it caught the committee's attention. 'Evidence has been brought before us', they wrote in their official report, 'that the use of the "oliver" . . . is unfit work for women or girls, with the exception of the "light" oliver used for making hobnails.'[2] Thus they recommended that women (and girls) be prohibited from using an oliver over a certain weight and from making chains exceeding a certain specified thickness.[3] Their suggestions then became the basis of various parliamentary proposals which, it was hoped, would be included in a new factory and workshop bill under consideration by the government. In February 1891 Lord Dunraven and Lord Thring introduced clauses prohibiting girls under sixteen from the trade and females from cutting iron over 1/4 inch or using an oliver if the hammer to be moved exceeded 4lbs in weight.[4] Paul Stanhope, MP for South Staffordshire and East Worcestershire, which included the nail and chain districts, entered his own proposals in the House of Commons in June of that same year. He wanted parliament to limit the size of iron women could work to make nails, spikes or rivets and to confine women's work in workshops within the hours of 6 a.m. and 6 p.m.[5]

These proposals to restrict women's work in this sweated trade led to an intensive and extensive extra-parliamentary debate. The Home Office was inundated with letters and met various deputations for and against the

---

[1] *House of Lords select committee on the sweating system: fifth report; with the proceedings of the committee, minutes of evidence, and appendix*, PP 1890, [c. 169] xvii. 287.
[2] Ibid. 301.
[3] Ibid.
[4] Lord Dunraven, who had served on the House of Lords Select Committee on the Sweating System, and Lord Thring introduced these clauses in a private bill unveiled in the House of Lords on 3 Feb. 1891. Their proposals were not included in a government bill which soon followed. Dunraven did move to add a clause along these lines to the latter on 13 July 1891.
[5] Debates of the House of Commons (cited hereinafter as *Hansard*), 3rd ser. ccciv. 354 (19 June 1891).

proposed measures. Working men continued in their efforts to secure limitations on the labour of their female competitors while employers and working women formed a strong lobby against them. Feminists appeared, once again, to defend women's right to work without undue state intervention. The press represented a new addition to the public dialogue and played a significant role by extensively covering the protest meetings and undertaking its own investigations of women making nails and chains. This chapter will examine these proceedings and suggest that they proved, unexpectedly, to be a decisive moment in the history of protective labour legislation. First, they illustrate the emergence of a change in the trajectory of the discussion on the dangers of women's work. The intensive discussion of women's nail and chain labour came to revolve around its physical dangers to them and their unborn children. Moreover, the government's attempt to resolve the sweating problem in this single trade led to the creation of a clause in the 1891 Factory and Workshop Act which opened the way for the new and more radical avenue of dangerous trades regulations.

Sweating in the nail and chain trade

The nail and chain trade was divided into a well-paid, highly organised and skilled male sector working in factories, and a sweated branch, predominantly female, located in domestic workshops attached to homes. Women comprised 48 per cent of the workers in the trade nationally and 68 per cent of the work force at its geographical centre in Staffordshire, commonly referred to as the Black Country. Long hours, poor working conditions and, above all, low pay, characterised women's work in this trade. The minimal skill needed and the almost complete lack of apprenticeship requirements, attracted women, children and men from the declining local industries, particularly miners and puddlers. The resulting oversupply of workers and competition between large numbers of middlemen were the major factors creating low wages. Desperate for work, people were at the mercy of those middlemen and subject to increasingly large wage cuts.

Male nail- and chainmakers had regarded competition from women working in domestic workshops as the cause of their low wages since the 1870s and frequently proposed the prohibition or restriction of their labour.[6] In 1875 working men unsuccessfully called for the prohibition or restriction of women's work before the government's royal commission. In 1883 they supported a bill to restrict the employment of females under fourteen in the

---

[6] They also proposed the registration of domestic workshops, enforcing uniform factory hours in them, and setting standardised price lists. For more on those proposals see Sheila Blackburn, 'Working-class attitudes to social reform: Black Country chainmakers and anti-sweating legislation, 1880–1930', *International Review of Social History* xxxiii (1988), 42–69.

trade while another attempt to limit women's work followed at the 1887 Trades Union Congress. At the latter meeting, Richard Juggins, secretary of the Midland Counties Trades Federation and National Nut & Bolts Makers Association, described the degrading way in which women worked in the trade. 'They worked', he said, 'in almost a nude state at the manufacture of chains, and only received for that labour 3s. to 3s. 6d. a week of sixty hours. The conditions under which they laboured were degrading and demoralising.'[7] The TUC passed his resolution to amend the existing labour laws to prevent the employment of women in the making of chains, nails, rivets, bolts or any such articles made of iron or steel because the work was unsuitable for them.

By the time the House of Lords Select Committee on the Sweating System investigated the trade in 1888, such efforts had made female chainmakers the centre of the sweating controversy. As a result, historian Sheila Blackburn has argued, they had transformed what was essentially an issue of wages into one of gender.[8] Indeed, this was the case when witnesses disclosed the horrible conditions of this labour for women before that important investigative body. In a manner reminiscent of the 1830s and 40s, several of them charged that work in the trade led to 'gross immorality and indecency' among the women. The Revd Harold Rylett, for example, testified that the small, unsupervised workshops were more conducive to immorality than the larger ones. Consequently, he preferred that married women should not work and that those who did did so in the 'safer' factories. Thomas Homer, secretary of the oldest union of domestic chainmakers, the Cradley Heath and District Chainmakers' Society, stated that in some shops 'where they have big young women blowing for them, very often there are doings that will not bear the daylight'.[9] Dire consequences also ensued at home for, he added, 'when night comes and they have both done their work there is neither fire nor comfort in the house; and very often it drives the man to go and get a pint of beer extra to what he would have'.[10] He was also sorry to report that many women had been trained from youth to make nails but were ignorant about domestic skills. If women stayed at home and fulfilled their proper duties, Homer concluded, there would be more comfort in working men's homes and their wages would rise. Both men made belated remarks about the physical impact of this work on women. Rylett was the source of the opening quotation of this chapter which emphasised that the use of the oliver led to the 'misplacement of the womb'. Homer offhandedly remarked on the decline in the physical condition of children born to chainmaking women. Such testimony illustrates an overlap of complaints about the evil effects of women's work; it led to immoral behaviour in the workshops, the neglect of their domestic duties

---

[7] *Annual report of the TUC* (1887), 42–3.
[8] Blackburn, 'Working-class attitudes', 52.
[9] *Select committee on the sweating system*, 286.
[10] Ibid. 270.

and it was injurious to their health. Both old and new lines of argument were presented to persuade the sweating committee that the state should restrict women's work in this trade.

The committee's response to these diverse comments was quite significant. They dismissed complaints about the immorality of women nail- and chainmakers saying that

> We found nothing whatever to justify these imputations on the character of the people. As a rule, they are well conducted, and although it may be objected that some of the work is unfit for women to do, there is no warrant for the assertion that it is indecent.[11]

To buttress their conclusion, they cited the testimony of a long-time resident and magistrate who testified that prostitution was wholly unknown in the district and that of the twenty-six illegitimate births recorded in the past half year, women chainmakers accounted for only three. Consequently, they dismissed the allegations of sexual misconduct but seized upon the few incidental comments suggesting the harmful effect of the oliver on women's reproductive functions. Thus, when they assessed the various alleged hazards of this work for women they were persuaded that it was unsuitable for them by the newer (or perhaps nascent is a better word) discourse of physical danger. And, as I have previously noted, various proposals followed from this committee's conclusion that the government should intervene to protect women from this harmful form of labour.

Fury over the forge

Employers and women workers protested against the proposals of Lord Dunraven and Lord Thring as both groups regarded them as a threat to their interests; women were a cheap source of labour for employers while women, in turn, needed to work and wanted to keep their jobs. Thus, both set out to prove that work in the nail and chain trade was neither unsuitable nor dangerous to women and that the women themselves were opposed to any restrictions. On 17 March George Green, secretary of the Employers Committee in the Nail and Chain Trade, sent to the Conservative Home Secretary, Henry Matthews, the details of their meeting in Birmingham protesting at the proposals. This meeting, representing twenty-three firms, unanimously resolved that if females were prohibited from making nails, rivets or chains, until they were sixteen years of age, the trade would practically be closed to them. Females, they asserted, 'can safely be allowed to use olivers up to 10 lbs. in weight without any injury to their health or spirits; that no female can make hob nails with an oliver less than 6 lbs. weight, and

[11] Ibid. 286.

lastly that females should not be interdicted from working iron up to 3/8 inch in thickness'.[12] That same day, the owner of Eliza Tinsley & Co. wrote a letter to the editor of a newspaper that made its way into the Home Office files. He did not object to general health and sanitary legislation but rather to laws which placed their trade in a special category. This legislation was particularly onerous, he wrote, because it was 'not actuated by any desire to benefit the females, but by the hope that in driving them from the making of nail and chains they would be securing more work and higher wages for the men'.[13] At a meeting on 7 April women testified to the amount of work they could do while employers dwelt on the hardships which would be inflicted on women if they were prevented from working in this trade.[14] When one employer claimed that men were taking an active interest in the matter because of women's competition, women responded, 'Let women work according to their strength.'[15]

Feminists, such as Millicent Garrett Fawcett, joined the campaign to block government interference with women's nail and chain work. On 28 March she wrote a letter to the editor of *The Times* protesting against the various proposals before parliament, proposals that she regarded as serving the interests of men in the trade. 'The men in the trade are', she wrote,

> in competition with the women, and every legislative restriction on women's labour tells to the supposed advantage of the men. The men can bring Parliamentary pressure to bear through their representatives; the women, whether employer or employed, cannot. As the men said to the women when a similar question arose a few years ago, 'We shall win in the end, because we have votes and you have not'.[16]

While not wanting to minimise the evils of low wages and long hours of work for women, she believed that the proposed legislation would only make matters worse by narrowing down their potential fields of employment.

Boucherett and the SPW also joined this defence of women's right to work by presenting evidence that the trade was not harmful to their health. Boucherett's friend, Helen Ogle Moore, conducted her own private inquiry among women working in the Black Country. When she had asked them if a 4lb. oliver was too heavy, they laughed and persuaded her to try it. 'She tried one of the ordinary ones', Boucherett related in a letter forwarded to the Home Secretary, 'and found it quite light, it went with a touch, a child of seven could have used it.'[17] Ogle Moore also learned that

---

[12] George Green to Matthews, 17 Mar. 1891, HO 45/9794/B5090E.
[13] CASW, 7 Mar. 1891, ibid.
[14] Green to Matthews (with report of this meeting and objections to the proposals), 9 Apr. 1891, HO 45/9794/B5090E.
[15] Ibid.
[16] *Times*, 28 Mar. 1891.
[17] Jessie Boucherett to Lord North (with a memorandum to the Grand Committee on Trade from the SPW), 20 Mar. 1891, forwarded to Matthews, HO 45/9794/B5090E.

at a place called Halesowen the women make railroad spikes and use olivers moving hammers of 18 lb weight. This work is hard and only strong women can do it. But I think the fact that strong women can and do use the olivers of 18 lbs. shows clearly that olivers of 8 lbs. are not and cannot be too heavy for ordinary women.[18]

Boucherett believed that it was a mistake for the government to interfere at all but if they found intervention advisable, the Home Office should raise the limit to at least 10 or 12lbs.

Ogle Moore joined Ada Heather Bigg of the SPW and MPs B. Hingley and Brooke Robinson at the most celebrated protest meeting of the whole campaign. A tremendous outdoor meeting held on 1 April 1891 had the *County Advertiser for Staffordshire and Worcestershire* writing that 'never in the history of the place had a meeting of such hubbub and noise been held as on that Wednesday night'.[19] It was, the reporter went on to write, a woman's meeting and whenever the male element attempted to question anything that was said the women refused to listen or hold their peace. Although a public meeting, women held the whip hand right through, the journalist continued, 'and the latter feeling they had the voting power with them on that particular occasion, were not slow to use it. They claimed that the men had no right to interfere in the proceedings, and demonstrated in unmistakable fashion accordingly'.[20] The meeting was billed as 'The voice of women on the question'.

'It was often said this or that occupation was unfit for women', Heather Bigg exclaimed before this excited crowd, 'but she had never been able to get a straight answer as to what it was that made work fit or unfit for women.'[21] Several years previously people had argued that pit-brow work was unsuitable for women and girls because it made them repulsive-looking. That same argument was used seven years ago against the nail- and chainmaking women, she added, to which a member of the crowd shouted 'They are the handsomest women living' and laughter ensued. Moreover, she addressed the crowd, 'When people argued the unfitness of an occupation for women because it involved physical exertion it must be asked – Was this industry exceptional? Was it more unhealthy than other occupations?'[22] Heather Bigg closed the meeting deprecating 'all legislation which will have the effect of driving women out of the nail and chain trades' and with a pledge to send a deputation of women nail- and chainmakers to meet the Home Secretary in London.[23]

On 17 April working women confronted Home Secretary Matthews as

---

[18] Ibid.
[19] *CASW*, 4 Apr. 1891, ibid.
[20] Ibid.
[21] Ibid.
[22] Ibid. She advanced similar arguments against proposed restrictions on women's work in the trade in 1883: 'Female labour in the nail trade', *Fortnightly Review* xxxix (1886), 827–32.
[23] Ibid.

part of a deputation led by Fawcett. Matthews told them that he had been informed by a certain medical officer that working with the heavier hammers was prejudicial to their health, especially those of child-bearing age. Immediately a very stalwart-looking woman exclaimed, 'I ha' had fourteen children, sir, and I never was better in my life.'[24] Matthews expressed polite satisfaction, and again quoted the doctor, whereupon all the nail- and chainmakers exclaimed in chorus, 'He's dead, sir!' as much as to say, 'He's dead and we are alive, so we needn't bother about him any more.'[25] The women told him that limiting them to the lighter hammers would restrict their wage-earning power and they resented that. At the end of this meeting, one woman turned to Fawcett and summed up their impression of Matthews: 'It's very 'ard upon the pore gentleman', she said, 'to 'ave to make the laws and not to know nothing about it.'[26]

Newspaper accounts of the deputation, including one in the most important paper in the district, the *Staffordshire Sentinel*, discussed the controversial subject of the effects of this work on the women's health. The subtitle of the article, 'Domestic service harder than chainmaking', suggested the comparative difficulties of this and other work for women. One woman testified that two of her daughters had found domestic service much harder than chainmaking. The paper was particularly impressed with a sixteen-year-old, Bertha Byng, who made spikes using an 18lb. oliver. 'A young woman of exceptional physique', the paper reported, 'she had been working for nine months and her health was better than ever.'[27] Its writer concluded that 'They were a sturdy set of women, neatly dressed and apparently none the worse for the work in which they are engaged.'[28]

The *Pall Mall Gazette* supported the deputation's main point: that women could work in this hard trade, enjoy good health and produce children. The issue of women's weakness, its writer argued, had been disproved by the paper's special representative who had made personal inquiries in the Black Country a few days prior to the deputation. His first inquiry took place in a chainmaker's shed where three women were employed. The group's 'boss', who was some months pregnant and wearing a 'defiant but not unpleasant expression', told him 'No, I don't believe this is this law they want. I say let each one do what they can. I like's the heavy work, and I don't see why I shouldn't be allowed to do it.'[29] When he asked her if this work hurt women more than men, she replied 'I don't know whether it hurts women more than

---

[24] There was no transcript of this deputation in the Home Office file but it was recounted in Ray Strachey, *The cause: a short history of the women's movement in Great Britain*, London 1928, 237.
[25] Ibid.
[26] Ibid.
[27] SS, 20 Apr. 1891.
[28] Ibid.
[29] PMG, 18 Apr. 1891.

men. . . . But I do know it doesn't hurt a woman a bit more to stand at a forge than to stand at the wash tub.'[30] Duly impressed with the strength and appearance of these women, he supported their right to work in the trade.

After the paper's photographer snapped pictures to accompany the story, the reporter 'cross-examined' some of the male workers. He concluded that 'They did not disguise the fact that the principle motive of the agitation was to prevent the lowering of the men's wages by the women's competition. The discovery of the injurious effects of the work upon the women's health was an afterthought.'[31] The reporter's perspective was reinforced with sketches of the male workers with the caption 'Male monopolists'. In the end the reporter objected to the special provisions currently under consideration because they establish 'the unjust principle that the freedom of women to earn their own living may be restricted by law in order to prevent women's competition with men, and in order to satisfy men's views of what is becoming and healthy for women'.[32] His visit to the Black Country convinced him that trade rivalry, rather than women's physical disabilities, was the driving force behind this campaign.

The divergent male and female perspective on women's work in the trade was further developed in the press. In an article published in the *Star* in March entitled 'Black Country slavery', William Price, of the Spike Nail Makers' Association, spoke about the conditions of life and work in the Black Country. He believed that women should not make nails at all and added that no man with any 'feeling of humanity' would desire to see women doing this type of 'grim and sordid' or 'degraded' toil. It was, he volunteered, his domestic creed 'that all married women should remain at home to mind their household affairs. As to the single women, there are hundreds of situations, and they may go into service'.[33] Admitting that it would be difficult to eliminate women's work entirely, he suggested limiting the size of nails which they could make. This would, he argued, 'place an embargo on toil which is too hard for them and ruinous to their health'.[34] Price reiterated his objections to this work because of its impact on women's health in a letter to the editor of the *Daily Chronicle*. He was astounded, he wrote, that one employer had never known a woman to be injured by the heavy work because, to his knowledge, 'there had been several cases of miscarriage and three or four deaths in my neighbourhood during the last six or seven years due to the unsuitable work done by women'.[35]

To further press their point, working men had sent their own deputation to Matthews on 9 April. Thomas Homer, president of the National

---

[30] Ibid. This article was also cited in the *CASW*, 25 Apr. 1891.
[31] Ibid.
[32] Ibid.
[33] *Star*, 23 Mar. 1891.
[34] Ibid.
[35] *DC*, 18 Apr. 1891.

Amalgamation of Chainmakers and Chain Strikers Association, who had testified before the sweating committee, told Matthews that his group had gauged the workers' feelings on the issue by a ballot. It showed that 1,719 workers favoured all women and young people starting to work at six in the morning and stopping work at six in the evening while 138 workers opposed that measure. Men, he explained, wanted a fixed time period within which women should work and did not want them to work at night. Overall, they had considered 6 a.m. to 6 p.m. as 'best suited to females' attendance to their Domestic duties'.[36] Furthermore, 1,746 workers thought that iron 1/4 in. thick was large enough for females to work while only 121 disagreed. Women and children can make 9/32 in. chain, Homer asserted, instead of the larger sizes and the men were willing to set the limit for heading spikes at 10/32 in. and 16/32 in. for sharpening.[37] If the government followed those recommendations, he told Matthews, it would be doing much to remedy the evils of the trade.

This ballot was immediately discredited by employers, who convinced the Home Office that trade rivalry motivated the entire campaign. Joseph Fellows, secretary of the Chain Masters Association, wrote to Matthews that the poll resulted from competition between the male and female workers in the district. 'I submit', he said, 'that the form, the mode of distribution and collection is so irregular, to say nothing about the opportunities for interference with the papers both during and after circulation, as to render any reported results, as altogether useless and unreliable.'[38] More significantly, he said, 'it must be borne in mind that this balloting has been wholly carried on by those men who are so strenuously opposed to women's labour'.[39] Working women, on the other hand, were almost unanimous in their opposition to special restrictions. The men were well aware, Fellows concluded, that by limiting the range of sizes at which women work, their wages would be reduced below the living wage, thereby driving them out of the trade.

The permanent under-secretary, Sir Godfrey Lushington, and two other Home Office officials, agreed that the ballot was suspicious. It purported to have been issued to male and females workers over fourteen years of age, he wrote, 'but having regard to the feelings displayed at the meeting of Female Operatives (see /8) it would appear to be very doubtful whether their "Ballot" can in the least be trusted'.[40] In view of the antagonism between the male and female operatives over the proposed restrictions, Lushington concluded,

---

[36] Notes of 9 Apr. 1891 deputation, introduced by Mr Stuart Wortley, Mr Brooke Robinson and Mr Stanley Hill, HO 45/9794/B5090E.
[37] Ibid.
[38] Joseph Fellows to Matthews, 11 Apr. 1891 (enclosing copy of the National Amalgamation of Chainmakers and Nail Strikers Association's ballot), ibid.
[39] Ibid.
[40] Minute signed by Godfrey Lushington, Malcolm J. Delevingne and E. G. (name unknown), 13 Apr. 1891, ibid.

female workers would not be likely to place any confidence in a plan for eliciting their views which emanated from the men's union.

Moreover, the Home Office considered other evidence and was particularly impressed by the information provided in defence of women's work. The minute on Boucherett's letter presenting information provided by Ogle Moore reads 'both speak of the women and girls as looking well and their views coincide with those expressed at the Master's meetings'.[41] The proceedings of the meeting at which Heather Bigg spoke, Malcolm J. Delevingne wrote, 'lend colour to the statements that the men are acting in their own interest in supporting the proposed restrictions'.[42]

It was against this backdrop that Stanhope introduced his measures on behalf of the male workers in his constituency. In their defence, he emphasised that the heavy nature of the work made it entirely unsuitable for women. He said that he was aware that 'there are a number of ladies, who are not acquainted with this work, and who object to the limitation of female labour in any way but the operatives (he included women) in the district are strongly in favour of some limitation'.[43] Moreover, he dismissed the supposed objections of women workers on the grounds that they 'had been put forward with a great deal of persistency by Gentlemen who are in favour of what I may call the unrestricted rights of women, and who dislike any interference with female labour, and I know that various meetings have been held in reference to this and other questions'.[44] He concluded by claiming that the workers' ballot (by now discredited), showed 'tolerably conclusively' the feelings of the operatives.

Stanhope's proposal to limit women's work in domestic workshops, would, Home Secretary Matthews said, 'run roughshod over a half a dozen valuable sections of the Factory Act applicable to industries all over the country'.[45] As for the heavy nature of the work, Matthews replied, 'The operatives distinctly asserted that there was nothing in the work that was the least oppressive, and they objected in the strongest manner to any limitation whatever, which they said they would resent as tyranny.'[46] These objections decisively defeated Stanhope's proposals; instead Matthews inserted a clause which became known as the 'dangerous trades' clause. It stipulated that

> Where the Secretary of State certifies that in his opinion any machinery or process used in a factory or workshop (other than domestic workshops) is dangerous or injurious to health or dangerous to life or limb, either generally or in the case of women, children, or any other class of persons, or that the provision for the admission of fresh air is not sufficient or that the provision for

---

[41] Delevingne minute, 25 Mar. 1891, ibid.
[42] Delevingne minute, 11 Apr. 1891, ibid.
[43] *Hansard*, 3rd ser. cccliv. 961 (19 June 1891).
[44] Ibid. 955–6.
[45] Ibid. 956.
[46] Ibid. 961.

means of escape from a factory in the case of fire is insufficient, the chief inspector may serve on the occupier of the factory or workshop a notice in writing either proposing special rules or requiring the adoption of such special measures as appear to the chief inspector to be reasonably practicable and to meet the necessities of the case.[47]

This general clause was intended to open up an avenue for dealing with individual trades carried out in especially hazardous conditions. The Home Secretary was empowered to certify a trade as dangerous whereupon the chief inspector of factories would propose special rules or the adoption of special measures. These rules or measures would then be submitted to employers and workers and, if their objections were cogent enough, the matter would be taken to arbitration.

The very public and contentious discussion of sweating in the nail and chain trade marks a significant transitional phase in the history of protective labour legislation. It included both new and old features with respect to its participants, their discourse and the government's course of action. There was a continuity with regard to the actions and attitudes of working men who, once again, pursued their own agenda while purporting to speak on behalf of women's interests. They did not testify to the long hours they themselves worked or to the insanitary conditions of their workplaces but rather to the dangers of the work for women and, in a few instances, the detrimental impact of women's work on their wages. Quite clearly trade rivalry was a key factor motivating them to lobby for the limitation of women's hours of work in domestic, and hence unregulated, workshops and the limitation of the kinds and sizes of nails and chains they could make.

Working men's arguments for the limitation of women's work in the nail and chain trade signifies the beginning of a shift in emphasis. The trade unionist argument that women's work impaired their ability to carry out their domestic duties and was 'unbecoming' or 'indecent', which harked back to the dominant discourses of previous short-time movements, was losing it force. Many contemporaries, most significantly Home Office officials, viewed such comments as thinly veiled attempts to bolster their weak position *vis-à-vis* their employers. Those rather transparent arguments were easily dismissed but incidental comments about women's health were, as I have suggested, duly noted by the House of Lords' sweating committee. When interviewed by the suspicious *Star* reporter, William Price defended his desire to limit women's work by stating that there had been several cases of miscarriages due to this 'unsuitable' labour. Another reporter, from the *Pall Mall Gazette*, had characterised such arguments as an afterthought.

The women who sponsored mass meetings and independently investigated women's work in the trade noted the mixture of arguments and shrewdly

---

[47] This clause was included in the Factory and Workshop Act, 1891, 54 Vict., IV. 159.

recognised the changing frame of reference. Heather Bigg, for example, questioned the very flexible criteria for labelling a trade unfit for women at the celebrated 1 April meeting. While the criteria were still in flux, Ogle Moore and Boucherett astutely downplayed their previous liberal rhetoric and couched their arguments in the newer terms of working women's physical condition. Moreover, they undertook their investigation specifically to prove that nail and chain women were healthy and that the trade posed no special harm to themselves or their ability to bear healthy children. Their first-hand evidence did have an important impact on state actions as the Home Office singled out, and stressed, their representations of the women's good looks and ability to work an oliver without injury to their reproductive health. Hereafter, as the public gaze repeatedly fell upon women workers, the effects of women's work on their reproductive abilities became the primary consideration. And, once again, women would gather and disseminate information about the health of the women workers under scrutiny in order to support or contest the consensus on the subject.

The discussion of women's labour in the nail and chain trade also illustrates the entry into the fray of new historical actors and the conspicuous absence of others. For instance, the press eagerly attached itself to the sexy and sensational story of thinly-clad women hammering at the forge in the Black Country. It also promoted its opinion on the subject and, like the SPW, challenged the idea that this work was especially dangerous for women. Their interest and efforts regarding nail and chain work in the Black Country marks the beginning of an increasingly important role for the press in the creation of protective labour legislation. The most significant absence was that of the WTUL, formerly known as the Women's Protective and Provident League, which had recently undergone a change in leadership and policy.[48] Under the earlier leadership of Emma Paterson, the group had vociferously opposed special legislation for women. In fact, the group's representatives at the TUC repeatedly spoke out against proposed restrictions on female labour in the nail and chain trade even as late as 1887.[49] As a result, they found themselves at odds with the trade unionists and allied with the SPW. However, that situation had dramatically changed within four years. Under the new leadership of Emilia Dilke and her niece, Gertrude Tuckwell, the group now embraced protective labour legislation. Several factors, Rosemary Feurer has argued, converged to create this major policy change: contact with working women

---

[48] For more on the history of trade unionism among women see Sarah Boston, *Women workers and the trade union movement*, London 1980; Sheila Lewenhak, *Women and trade unions: an outline history of women in the British trade union movement*, London 1977; Theresa Olcott, 'Dead centre: the women's trade union movement in London, 1874–1914', *The London Journal* ii (1976), 34–50; Norbert Soldon, *Women in British trade unions, 1874–1976*, Dublin 1978; and Deborah Thom, 'The bundle of sticks: women, trade unionists, and collective organization before 1918', in John, *Unequal opportunities*, 261–89.

[49] In this instance Clementina Black opposed the suggested limitation from Richard Juggins: *Annual report of the TUC* (1887), 43.

who wanted state aid, discouragement over the ability to organise women, the development of 'trade union consciousness' among some of its members and the influx of supporters of socialism, whose views on state intervention were diametrically opposed to the classical liberalism of earlier feminists.[50] Thus, they mounted no opposition to the possibility of special regulations for women's work in this sweated trade.

The WTUL's change in policy meant a split within the women's movement over protective labour legislation. For, while it had modified its views, the women in the SPW maintained the position they had taken up during the 1870s. They continued to oppose what they considered to be discriminatory legislation passed to serve working men's interests. As this chapter has shown, they protested against the specific measures intended for the nail and chain trade and any general measures that would restrict women's work in domestic workshops and homes. Thus Boucherett wrote in 1891:

> Let the reader carefully note this invasion of the home of our cottage workers is done with the avowed object of driving work out of the homes into the factories. This would not lead to so much change in the case of men, whose home industries are comparatively few, but for women it means little short of a revolution.[51]

Henceforth, women's groups were pitted against each other as they tried to show that they represented the true opinion of working women on the divisive issue of special labour laws for them. This rupture and reorientation has led historians to apply the labels 'social feminists' and 'equal-rights feminists' to distinguish between members of the WTUL and the SPW.

I want to suggest that this episode exhibits the outlines of a pattern that would be repeated in the decades to follow. The impetus and initiative for legislation would not be found in parliament but in the extra-parliamentary arena. In this instance, working men in the trade brought the issue of women's work to national attention during the House of Lords' sweating investigation and prompted various MPs to include proposals for its regulation. Women's groups and the press, meanwhile, were not satisfied with government inquiries and undertook their own. First-hand evidence, such as that collected and presented to the Home Office by the SPW, played a significant role in persuading the government that women workers were healthy.

In the end, the government's solution to this labour problem was the creation of the dangerous trades clause. This was a truly innovative and significant consequence of the public discussion of sweating in the nail and chain trade. For, as the next chapter will illustrate, it became the basis for dangerous trades regulations.

---

50 See Feurer, 'The meaning of "sisterhood" ', 233–60.
51 Jessie Boucherett, 'The new factory legislation', ER xxii (1891), 75.

3

Dangerous Trades Regulations and the White Lead Trade

Home Office officials quickly urged the Home Secretary to exercise the new power conferred upon him through the dangerous trades clause of the 1891 act and, yet, they did not refer to the nail and chain trade. Instead, on 26 April 1892, the chief inspector of factories, R. E. Sprague Oram, requested an order to declare the white lead trade dangerous.[1] Following its official designation as a dangerous trade in May, a series of rules were drawn up to supplement those that, it should be noted, had been put in place by the 1883 White Lead Act.[2] Thus, as it stood in 1892, employers were required to provide bathing accommodations for men and women, a sufficient supply of hot and cold water, soap, brushes and towels, overalls and respirators, and a supply of an acidulated drink for the workers. They were also required to ensure that every man and woman took a bath once a week and did not leave the lead works unless properly cleansed from the lead. For their part, workers were required to wear overalls and respirators and instructed to pay careful attention to personal hygiene. Finally, a doctor was required to examine each worker every week while those who were taken ill were supposed to report to him. In addition, May Abraham was appointed in 1892 to investigate the conditions of women's white lead work on behalf of the Royal Commission on Labour. The government was pleased with its actions and thought that it dealt sufficiently with the hazards of this particular trade.

The government's sense of accomplishment, however, was undermined in December 1892 when the *Daily Chronicle* published a series of sensational articles cataloguing the horrific conditions in which women worked and died in the trade and, especially, the high incidence of death among their infants. The paper succeeded in attracting massive public attention and in creating an industrial scandal. This chapter will examine the course of events following the exposure of the special dangers of this work for women and the government's resolution of the problem. Herbert Asquith, who became the Liberal Home Secretary in August 1892, included a further dangerous trades clause in bills to amend the 1891 Factory and Workshop Act in 1894 and

---

[1] Oram to Lushington, 26 Apr. 1892, HO 45/9856/B12393A.
[2] The provisions of the 1883 White Lead Act are outlined in B. L. Hutchins and Amy Harrison, *A history of factory legislation*, London 1903; repr. London 1970, 202.

1895.³ The resulting legislation bestowed unprecedented power upon the Home Secretary by allowing him to restrict or prohibit the labour of certain persons, meaning women and children, of course, from parts of or entire trades considered especially dangerous to them. The consequences for women workers in the white lead trade were immense as they were eliminated from the most dangerous, and highest-paying, portion of the trade in June 1898. In so doing, the English government created legislation which resembles late twentieth-century American 'foetal protection' policies.

Investigating the white lead trade

Contemporary investigations revealed that more women than men were employed in this trade and that they worked in its most dangerous sectors.⁴ In Newcastle-upon-Tyne, where the five major white lead firms in England were located, 565 females and 328 males were employed in 1896, 571 females and 329 males in 1897.⁵ Women worked in the 'blue beds' which consisted of small rows of small earthenware pots filled with acetic acid. There, a group of five women would place small lead cakes on top of the pots, surround them with bark, put a plank over the pots, and build them into stacks. After three months they placed the pots in an oven to dry for a fortnight. After the interaction of heat and acetic acid had produced a cake of white lead, the lead was then placed on a roller. Water was poured over the rollers which crushed the lead cake after which the water and lead ran through a series of tanks called washbecks. The lead was then placed in an oven to dry and then removed by women, working in what were called the 'white beds'. The removal of the dry lead was the most dangerous part of the process because lead dust flew around and if it entered a worker's system, through the nose or mouth, lead poisoning could result.

Women earned high wages for their work in this hazardous trade. According to the factory inspector for Newcastle-upon-Tyne, the average wage for women was between 2s. and 2s. 6d. per shift. If women worked two

---

³ The clause first appeared in his 1894 bill, introduced in parliament on 30 Apr. and withdrawn on 18 July. The second bill, which included the same clause, was introduced on 1 Mar. of the following year and was approved on 23 July.

⁴ In addition to government reports see Mrs Charles Mallet, *Dangerous trades for women*, London 1893, and Helen Ogle Moore and Edith Hare, 'Report to the Society for Promoting the Employment of Women on the work of women in the white lead trade, at Newcastle-upon-Tyne', in Jessie Boucherett and Helen Blackburn (eds), *The condition of working women under the Factory Acts*, London 1896, 77–84. For more on the history of the trade see D. J. Rowe, *Lead manufacturing in Britain: a history*, London 1983.

⁵ These statistics, provided by Factory Inspector H. J. Wilson, were reprinted in Dr Thomas Oliver, 'Lead and its compounds', in Dr Thomas Oliver (ed.), *Dangerous trades: the historical, social, and legal aspects of industrial occupations as affecting health, by a number of experts*, London 1902, 282–372 at p. 299.

shifts a day, and many did, they could earn £1 or more per week.[6] These were good earnings considering, as the 1886 wage census of the Board of Trade revealed, the average weekly wage for women was 12s. 8d.[7] One contemporary noted that this trade 'represents for women very much what the Dockers' industry is to men', meaning that it attracted the poorest and roughest class of women.[8] This perspective was seconded by one senior factory inspector who wrote that the female workers were generally 'young girls who are particularly without the comforts of a good house, many of them live a questionable life, they expose themselves to cold and are frequently in a state of chronic starvation before going to the lead works and therefore in a fit state for rapid breaking down under the influence of lead'.[9] Both highlight the critical facts that many of female employees in the white lead trade could not find work elsewhere and took up this dangerous work because of poverty. Lacking adequate food and shelter, many were already physically unfit as they began work and were thus predisposed to illness.

May Abraham's report for the Royal Commission on Labour substantiated the dangerous nature of the trade for women. Between 1883 and 1888 a total of 135 cases of lead poisoning were admitted to the Newcastle Royal Infirmary; ninety-four were women and forty-one men. Five of the women and three of the men died. Moreover, inquests in the Newcastle district between 1889 and 1892 further showed that twenty-two women and one man died of lead poisoning; the majority of those deaths occurred among workers aged between seventeen and thirty. Abraham based a substantial portion of her report upon her interviews with women currently suffering from lead poisoning. The case studies were horrible. For example, one worker aged sixty-three, who had worked in the rollers, stoves and white beds, had had her first attack six weeks after beginning work; she was ill for four months and spent time in a workhouse. A second worker, who had worked in the white beds and rollers was blind and paralysed. 'A.B. Optic neuritis. Age 28 years', Abraham wrote, 'continued to work off and on for two years, during which she had frequent attacks of lead colic, "wrist drop" and epileptic fits. She was then attacked by optic neuritis, and is now in the workhouse totally and permanently blind.'[10] Finally, another woman had been sick three times while she worked in lead. She had six children; three died at an early age of

---

[6] Report, Richard Johnson to Whately Cooke-Taylor, forwarded to Oram, 24 Feb. 1896, HO 45/9856/B12393AC.
[7] George Wood, 'The course of women's wages during the nineteenth century', in Hutchins and Harrison, *A history of factory legislation*, appendix a, 257–308 at p. 261.
[8] Mallet, *Dangerous trades for women*, 8.
[9] Report, Arthur Henderson to Oram, 29 Dec. 1892, HO 45/9848/B12393A.
[10] *Royal Commission on Labour: the employment of women, reports by Miss Eliza Orme, Miss Clara E. Collet, Miss May Abraham and Miss Margaret Irwin, lady assistant commissioners, on the conditions of work in various industries in England, Wales, Scotland, and Ireland*, PP 1893–4, [c. 6894–xxiii], xxxvii. 51.

convulsions while the other three, born during intervals of abstention from work, were healthy.

Abraham concluded her report with a very important section entitled the 'Special susceptibility of women to lead poisoning'. She extensively quoted Dr Thomas Oliver, Professor of Physiology at the University of Durham and physician to the Newcastle Royal Infirmary, was who considered the expert on lead poisoning. He maintained that women were more susceptible to the disease than men and that they suffered from its ill-effects at an earlier age and with more severity. Moreover, she noted, he placed great emphasis upon the evil effects of lead on the offspring of women lead workers. These opinions were also voiced by employers such as Mr Slaen, the manager of the Mersey White Lead Manufacturing Company, who told Abraham that he had discontinued the employment of women because of the greater incidence of illness among them. They absorbed more lead, he said, 'owing to the nature of their clothes, which collect dust more than men's clothes do'.[11] He also objected to the employment of women because of the 'inevitable injury to their children'.[12] Abraham concluded that the 'unsuitability of women's dress might easily be altered by regulations, but in further information received from Dr Oliver special disabilities are suggested which appear to be irremovable'.[13]

'Death in the workshop': poisoned women and infant slaughter

The Home Office received Abraham's report just a few weeks before the publication of two articles in the *Daily Chronicle* on 15 and 21 December entitled, 'Death in the workshop: white cemeteries: how women are poisoned' and 'Death in the workshop: white cemeteries: massacre of innocents'. They were special investigations of women working in the 'white beds' of the trade where they removed the dry white lead from the ovens. The first article began with an explanation of the article's subtitle, 'White cemeteries', which, the journalist wrote, 'is a name by which the white lead works are known amongst the women here whose fate it is to spend their lives, and very often to meet their death, in the manufacture of one of the most deadly of poisons'.[14] He provided his readers with a sketch of the 'type' of woman who undertook this dangerous but high-paying work. She might be the elder daughter in a family where the father was unemployed, a widow with a family to support, a wife whose husband was a loafer, or a girl whose character would not stand scrutiny. 'By a fatal process of selection', he wrote,

[11] Ibid. 53.
[12] Ibid.
[13] Ibid.
[14] *DC*, 15 Dec. 1892, HO 45/9848/B12393A.

these women go to the works ripe for death, their systems bared for the lead by poverty and impaired vitality. The course of labour which makes them carry and hoist tons of dead weight on their heads, which lashes their loaded bodies backwards and forwards, as no slave-master's whip, but only the dire need and agony of your hungry ones could – this is the best training for the death by poison which lies ready on every side, trodden into powder under foot[15]

He graphically described the white lead powder invading the women's bodies, 'gaining access under all coverings to every part of the body, gasped into the lungs and swallowed in the saliva, and absorbed through the skin'.[16]

Under a section entitled 'Paralysis, convulsions, and death', the author outlined the horrible progression of this disease. A woman would become anaemic, have headaches and vision problems. This would be followed by convulsions and death. If, and he stressed if, a woman survived she would continually suffer from colic, vision problems and paralysis. He personalised this tragic sequence of events by recounting the story of seventeen-year-old Charlotte Elizabeth Rafferty, as told to him by her parents the day after her funeral. Each day of the five months she worked in the lead factory, she came home and cleaned herself and her underclothing of the lead dust which resembled flour. Her father recalled that he used to tell her to 'Go and shake 'em out at the back door, and don't smother us with it.'[17] Her mother showed the journalist her skirt which was no longer dusty but stained with white patches. Charlotte, they told him, had never been ill a day in her life before she went to the lead works. They continued:

> She had not complained up to the week before, when she got very sleepy of an evening, and they told her to go and stand at the front door and to freshen herself up. Her spirits were first rate, and she roused up and sang 'Maggy Murphy' the evening before she went to the works for the last time.[18]

She subsequently had 'a fit' and never regained consciousness before her death, attributable to lead poisoning.

The journalist also introduced the public to Dr Oliver whose testimony grimly underlined the gendered dimension of the problem. The writer noted that 'Dr Oliver is most positive as to the far greater liability of women to lead poisoning than men.'[19] And, he quickly added, 'he has certainly convinced me that there [are] reasons – apart altogether from the question of maternity – which makes the employment of women in this work a fearfully grave responsibility for the state to undertake, looking at it from the purely physiological standpoint'.[20]

15 Ibid.
16 Ibid.
17 Ibid.
18 Ibid.
19 Ibid.
20 Ibid.

This last line of inquiry was further developed in the second article, subtitled 'Massacre of the innocents'. 'Does the Home Office', the writer asked, 'know that the children of the white lead mothers are, as a rule, doomed before they are born?'[21] He substantiated Oliver's opinion about the dangers of this labour by describing its impact upon unborn children. 'Dr Oliver's contention', the author asserted,

> that the normal thing is for a child of the white-lead woman to die soon after birth is borne out by the following cases which have come under his notice. The liability to miscarriage, which is one of the penalties of the lead trade, is also brought out.[22]

According to the four case studies presented, out of a total of thirty-three possible children only seven survived infancy, sixteen died of convulsions shortly after birth, while the women suffered a total of ten miscarriages. After reviewing the existing regulations for this dangerous trade, the author concluded that the government licensed 'infanticide'.

Lastly, the journalist was critical of the special rules which, he contended, 'grant immunity to the capitalist for his murderous process, and throw the defence upon the workpeople'.[23] While mentioning the class dimension of this issue, he was ultimately more concerned about its gender component. He was especially astounded that the government did nothing to prevent at-risk mothers from working in the deadly lead. He advised an immediate government inquiry into the production of a harmless white lead as 'it has paid dearly enough in human life for the neglect of scientific advice'.[24] In the meantime, the writer concluded that the prohibition of women would be the best course of action considering they were 'the weakest and most helpless people whose destruction is heeded no more than that of the babes who go down with them before this frightful scourge'.[25] As for the unemployed women, he believed that sympathetic people would help find alternative, and more suitable, jobs for them.

These alarming articles, a mixture of highly sensational statements and Oliver's grim medical conclusions, were the catalyst to intensive public and governmental interest in more stringent measures to protect women employed in this highly dangerous trade. And, very significantly, they set the terms of the discourse in the ensuing debate. For example, on 7 January 1893 the *British Medical Journal* published an article praising the *Daily Chronicle* reporter for 'doing his best to arouse public opinion to the waste of life that is going on in certain of our industries'.[26] The article endorsed Oliver's opinion

---

[21] Ibid. 21 Dec. 1892.
[22] Ibid.
[23] Ibid.
[24] Ibid.
[25] Ibid.
[26] 'Dangerous industries', *BMJ*, 7 Jan. 1893, 7.

as to the greater susceptibility of women to lead poisoning and argued that the Home Office should ban them from the trade. Questions in parliament also followed. On 17 February 1893 one MP drew the Home Secretary's attention

> to the reports in the *Daily Chronicle* newspaper of the 15th, 21st, and 28 Dec., to the paper on 'Deadly Poisonous Trades' in the *Fortnightly Review* of Feb. 1893, wherein the writer alleges that 64 deaths from lead poisoning occurred in one factory alone in six months, and to the article entitled 'Wanton Waste of Life' in the *Whitehall Review* of the 4th Feb. and asked what steps the Government intended to take to investigate, and if possible, prevent the loss of life in white lead factories.[27]

The *Daily Chronicle* articles also prompted Chief Inspector of Factories Oram to elicit comments from the superintending inspector of Scotland and Northern England, Arthur Henderson. Henderson, who had nearly twenty-five years experience with the white lead trade, countered, 'It is an exaggerated and highly coloured statement of the case and contains just a sufficiency of truth to save it from condemnation as a piece of fiction.'[28] Lead poisoning, he wrote, affects a small minority of women, but the article leads one to believe that all women get poisoned. He quoted Oliver's opinion to make his point that

> It is the weak, the careless, the ill cared for, and the dissipated who suffer chiefly, but it is a libel on many hundreds of honest, industrious, and responsible women both married and single, to say that these and such as these constitute the majority of female workers in the white lead factories.[29]

Those, he added, who were careful in regard to personal cleanliness and were well cared for at home hardly ever suffered. Finally, Henderson added his opinion that 'as with most other diseases there was not only an individual but a family predisposition to lead poisoning. Whole families will readily suffer, others scarcely at all – this observation applies to both sexes'.[30] He personally believed that the government's special rules were sufficient precautions against lead poisoning.

Despite Henderson's encouraging report, Oram felt that the considerable publicity surrounding women's work in the trade mandated government action. 'Considering', Oram wrote to Home Secretary Asquith, 'the interest recently manifested in the question of the manufacture of white lead (which is evidenced by articles in the Newspapers and Reviews)', a departmental committee should undertake an extended inquiry into the subject.[31] Asquith

---

[27] *Hansard*, 4th ser. viii. 1704 (17 Feb. 1893).
[28] Report, Henderson to Oram, HO 45/9848/B12393A, 1.
[29] Ibid. 4.
[30] Ibid. 4–5.
[31] Oram to Lushington, 9 Feb. 1893, ibid.

readily agreed and a special departmental committee to investigate the conditions of work in the white lead trade met in late 1893.[32] It questioned workers, foremen, employers, chemists and a few working women. In view of the tremendous concern about women working in the trade, it focused a great deal of attention on the issue of their alleged greater susceptibility to lead poisoning and its effect on their offspring. And, it should be noted, the committee entered its investigation already considering the possibility of eliminating women from the trade.

Dr Oliver, the White Lead Committee's medical expert, interviewed thirteen doctors, eight of whom endorsed his theory of women's greater disposition to lead poisoning. None, however, explained the exact reason for that fact but they could, and did, readily talk about the detrimental effect of lead on the children of lead workers. In contrast to medical testimony, several working women who had been in the trade for many years, testified as to their good health and that of their children while several doctors and managers suggested circumstantial causes of the disease. One doctor noted that 'there is not doubt that more women than men suffer, but that I should think, is partly due to the nature of the employment, and that a larger number of women are exposed to the influence'.[33] Other factors mentioned were the conditions of the particular factories, the degree to which the government's special rules were implemented, and the cleanliness and personal habits of the workers, their nourishment and constitutions.

After hearing the testimony, the committee concluded that more cases of lead poisoning occurred where precautions were not taken, that women were more susceptible to illness than men and that this trade was not a fit occupation for them. It recommended that no women under eighteen be employed in the trade and that 'no women be employed in white beds, the rollers, the washbecks, the stoves, or in packing white lead, and that these departments should consequently be worked in future only by adult males'.[34] Medical opinion was clearly a decisive factor in the decision to make this drastic recommendation.

Following the publication of this report in 1894 and more newspaper reports of women dying in the trade, Asquith faced questions in parliament about implementing the committee's recommendation. John Burns, a prominent Socialist labour leader, was one of many politicians inquiring about the government's course of action. On 26 July 1894 he asked whether

---

[32] This committee included Henry James Cameron, factory inspector for London, Dr Thomas Oliver, Arthur Pulliam Laurie, Fellow of King's College, Cambridge, A. Dupre, lecturer on chemistry at Westminster Hospital Medical School, and Henry J. Tennant, assistant private secretary to the secretary of state and secretary of the committee.

[33] *Report of the departmental committee appointed to inquire into the conditions of labour in the various lead industries, into the dangers to the workpeople employed therein, and to propose remedies; with evidence, appendices, and index*, PP 1893–4, [c.7239] xvii. 792.

[34] Ibid. 25.

considering the extraordinary susceptibility of girls and women to complaints arising out of this occupation, the Right Hon. Gentleman has considered the advisability of carrying out the recommendations of some of his Inspectors and medical experts, and excluding girls and women from this particular employment?[35]

Asquith replied that he had considered such a course of action but lacked the statutory power to take it. He sought to remedy this situation with a new dangerous trades clause included in a new factory and workshop bill introduced, but dropped, in 1894, and reintroduced in 1895.

Asquith's bill contained a dangerous trades clause to empower the Home Secretary to prohibit the employment or limit the period of employment of any class of workers engaged in a dangerous trade. 'He had found', he said, 'in framing these special rules that in connection with the lead industries and other dangerous trades he had been very much hampered by not having the power to prohibit altogether certain classes of people from engaging in certain processes.'[36] By any class of workers he meant, of course, women and children. The lack of such power had prevented the enactment of what he saw as the most valuable recommendation of the White Lead Committee: the prohibition of women from certain processes in the white lead trade. He was certain that this exceptional measure was essential to the health and life of the workers and commented that 'the powers were not at all likely to be abused in any sense that was oppressive or injurious to the liberty of the trade'.[37]

The extra-parliamentary debate

The possibility of extending the Home Secretary's power over women's work in dangerous trades elicited a tremendous response. Asquith heard from women, and only secondarily, men in the trade union movement as well as women's rights groups. As in 1891, the latter groups persisted in their opposition to special regulations for women while the women in the labour movement demanded more stringent regulations. The Home Office, then, was pressured by two major lobbies vying to show that they represented the true opinions of working women. They expressed their opinions in 1894 and 1895 through deputations to the Home Office, meetings and articles in newspapers and journals. They produced investigations and offered their opinion and, in so doing, they emerged as key participants in the extra-parliamentary discussion of Asquith's dangerous trades proposal.

---

35 *Hansard*, 4th ser. xxvii. 1022 (16 July 1894).
36 Ibid. 4th ser. xxxi. 178 (1 Mar. 1895).
37 Ibid.

The Women's Industrial Defence Committee, formed in 1892 to safeguard women from protective labour legislation, viewed the proposed clause as yet another way of handicapping women in the labour market and raised that objection at deputations to Asquith in 1894 and 1895.[38] Eleanor Whyte, a bookbinder who served as the group's honorary secretary, opened the 1894 deputation by claiming that the entire bill before parliament was an attempt to treat women like children. And, she emphasised, women were especially fearful of the possibility of extending the power of male officials over women workers through its dangerous trades proposal. Acutely aware that trade rivalry had often led working men to press for limitation on women's work, she and her group believed that this provision could be used to eliminate unwelcome women from the labour market. Thus she told him:

> When I went upon a deputation with regard to the chain and nail makers we had an official report to state that the trade was not unhealthy. I know the trades that are unhealthy and the work that is injurious to women. It is just those trades and work in which men think that women interfere with them. ... We want to work and we do not want to be restricted.[39]

Such interference, veteran anti-protection campaigner Boucherett asserted, would be harmful to the women workers who could lose their jobs. 'These women who are to be turned out of the lead works', she told the government officials, 'have been twenty and thirty years in the trade and have never suffered and are now hale hearty women still earning good wages.'[40] 'Even in the lead trade', she added, 'when the employers provide food, milk, oatmeal, and broth for the workers the women do not suffer from the lead.'[41] She contended that unemployment would pose greater hardships for these recipients of state aid than their present employment in the white lead trade. Similar arguments were put forward in a second deputation in 1895.[42]

Home Office officials attempted to present a positive portrayal of protection at both meetings. In 1894 Asquith flatly denied the charge that women were treated like children under protective labour legislation. That argument, he pronounced, had been used a lot recently but:

> The whole foundation of that Legislation rested upon the assumption that public safety and public interest required that women should be treated in some matters, and among others in the hours of employment, in a different footing from men; and an argument based upon the assumption that there was no distinction between the two sexes and that they ought to be treated alike

---

[38] The SPW formed this special subcommittee: 'Thirty-third annual report of the Society for Promoting the Employment of Women', ER xxiii (1892), 182–3.
[39] Transcript of deputation from the WIDC, 26 June 1894, HO 45/9881/B16265, 25.
[40] Ibid. 67.
[41] Ibid.
[42] See transcript of deputation from the WIDC, 20 May 1895, HO 45/9890/B17300.

was an argument fatal, not only to the Bill now before Parliament, but to the whole series of Factory Acts from the very beginning.[43]

For him, and many others, protection was both natural and necessary because of the differences between the sexes. He added further weight to the argument for protection by telling them that 'There are kinds of industries which are peculiarly injurious to them – far more injurious to them than they would be to men, and as to which they cannot take the same precautions.'[44] He tried to allay their fears with the assurances that he was sympathetic to the plight of female workers and that the government would only pass laws 'necessary for the safeguarding of the health and lives of the female and infant children workers'.[45] At their 1895 encounter, George Russell, under-secretary for the Home Office, repudiated the charge that the government was going to use its new power to serve men's interests by 'shoving' women out of trades. They only wished, he remarked, to protect them 'from the dangers to which you are exposed in the pursuance of your calling'.[46]

Members of the SPW undertook two activities in order to counteract the sensational reports of illness and death in the trade and to prove that further legislation was unnecessary. First, they emphasised neglected information contained in the government investigations to show that factors other than the sex of the worker contributed to the development of the disease. Their leader, Boucherett, focused on information gathered by Abraham, Eliza Orme and Margaret H. Irwin for the 1892 Royal Commission on Labour. Abraham had reported death and illness in a white lead factory in Newcastle where precautions were 'very imperfectly observed'.[47] In contrast, Orme and Irwin reported no deaths and only an occasional case of colic at the factory they visited in Scotland. The medical officer at that factory had told them that colic

> mostly occurs where the general surroundings are bad, causing impairment of the general health, thereby predisposing to illness, but as a rule we have a healthy lot of girls and women to deal with, who are rarely ill, although many of them have been employed in the factory a great number of years.[48]

He concluded that where proper precautions were taken workers ought to have almost complete immunity from the bad effect of lead.

In addition, in 1895 two members of the SPW became investigators and revisited the five factories in Newcastle-upon-Tyne previously investigated. Helen Ogle Moore and Edith Hare intended to prove that the press and

---

[43] The Home Secretary's response was published in *The Times*, 27 June 1894.
[44] Ibid.
[45] Ibid.
[46] Transcript of deputation from the WIDC, 20 May 1895, 12.
[47] Jessie Boucherett, 'Lead works and some other unhealthy industries', *ER* xxv (1894), 10.
[48] Ibid. 12.

medical men had created a distorted picture of the causes and incidence of lead poisoning. Their view that circumstantial factors led to the development of the disease was validated by Dr Redmayne of the Royal Infirmary who told them that the evils of lead could, in most cases, be avoided if workers were careful and clean. To destroy the newspaper impression that all women and their unborn children perished, they interviewed women who had been employed for twenty or thirty years without illness.[49]

While they did not discount the dangers of this trade to women and their children, women in the SPW and WIDC argued that prohibition was not necessary and, in fact, would inflict greater hardships on women in the trade. Since they believed that conditions in the lead works caused illness, they argued that the dangers could be effectively diminished if the government's special rules were more stringently enforced. Moreover, they charged that prohibition would inflict tremendous suffering upon these women who earned good wages. The workers interviewed by Ogle Moore and Hare spoke of the work 'as very hard for the time but the hours are comparatively short; they knew of the dangers of the stoves and white beds, if they were not careful, but they asked what would become of them if it was stopped – work was scarce; they could get nothing else'.[50] After losing their jobs, Boucherett argued, they would be driven into other overcrowded, low-paying jobs, the workhouse or worse, meaning prostitution.[51] Moreover, her organisation estimated that the number of women who would lose their jobs would be three times the government's estimate of 600.[52]

The Women's Liberal Federation and the Women's Emancipation Union were new organisations that entered the crusade against special legislation for women.[53] Eschewing the argument that sexual difference mandated and justified special laws for women, they made the radical claim that dangerous trades measures should apply to both sexes. While the WLF approved of the general provisions of the bill at their 1895 council meeting, they 'viewed with apprehension those clauses which proposed restrictions on women only,

---

[49] Ogle Moore and Hare, 'Report to the Society for Promoting the Employment of Women', 79.
[50] Ibid.
[51] Boucherett made this point in the deputation to the Home Secretary in June 1894.
[52] Memorial, SPW to Home Secretary Herbert Asquith, Jan. 1895, HO 45/9848/B12393E.
[53] Helen Gladstone, wife of William Gladstone, the countess of Aberdeen and the countess of Carlisle were among the early presidents of the former group founded in 1886. It supported the implementation of Liberal principles in legislation and, as they phrased it, just legislation for women. According to the 1895 *Annual report of the executive committee*, it had a membership of 82,000 women in 448 branches across the country.

The WEU, an association formed to secure the political, social and economic independence of women, was formed in 1892. John Boyd Kinnear, Mona Caird and Elizabeth Wolstenholme Elmy, three prominent and long-standing women's rights activists, were among its members. This group fought for equality between men and women in industry through equal freedom of choice in career. For more on this group see WEU, *The Women's Emancipation Union: its origins and its work*, Manchester 1892.

which will, therefore increase the existing disadvantages under which women labour'.[54] More specifically, as Annie Veness, honorary secretary of their Swindon branch wrote to the Home Office, men should also benefit from the dangerous trades regulations.[55] At its annual conference in 1894, the WEU passed a resolution protesting 'on economic, social, and moral grounds against the imposition of special sex restrictions upon the labour as injurious to the best interests of men and women alike, and claims that the legal protection for workers be equal for both sexes'.[56]

Finally, the prospect of closing this trade to women was vehemently opposed by those middle-class women who were fighting for more work opportunities for women and by the workers who were compelled by economic necessity to undertake this difficult work. Working women, whose opinions had not been solicited, made personal pleas to Asquith. For example, on 10 May 1895, a deputation of the Women's Employment Association, comprised of lead and Staffordshire pottery workers, expressed their apprehension regarding the proposed power of the Home Secretary. They asserted that they were healthy and that conditions had improved since the enactment of the special rules.[57] In June 1896 the Home Office received a petition signed by 524 lead workers from the Newcastle-upon-Tyne area asking it to consider the impact that a ban on their labour would have upon them. 'Many of us', they wrote, 'are widows with large families to support, others have no other means of getting a living except by this kind of work.'[58] These necessitous women did not want 'protection' from the state, since it would entail the loss of their livelihood.

The Home Office quickly and decisively dismissed all the contentious arguments made by the various women's organisations. The suggestion that men receive equal protection led to many dismissive remarks from Home Office officials. This is not surprising since these women essentially advocated the destruction of the basic premise of protection, the premise that men were free agents who could look after their own interests while women could not. In 1894 Chief Inspector of Factories Oram wrote that he did not think it advisable to extend this type of protection to men while his associate, Godfrey Lushington, wrote more bluntly 'I disagree from all the suggestions made by these ladies.'[59] Another official wrote in an exasperated tone: 'To prohibit the employment of men as of women and children in any process would of course be to prohibit the process. But I suppose the Women's Rights Societies will object to a restriction of women's labour which does not apply

---

[54] This resolution from the WLF was sent in a letter to Asquith, 17 Apr. 1895, HO 45/9889/B17300.
[55] Annie Veness, Swindon WLF, to Asquith, 19 June 1894, HO 45/9881/B16265.
[56] *Third report of the Women's Emancipation Union* (1896), 30.
[57] The proceedings of this deputation were reported in the SS, 17 May 1895.
[58] Petition, 9 June 1896, HO 45/9856/B12393AC.
[59] Oram minute, 25 June 1894; Lushington minute, 3 July 1894, HO 45/9881/B16265.

to men.'⁶⁰ Moreover, Dr Benjamin A. Whitelegge, who had become the new chief inspector of factories in 1896, commented on the worker's petition that they were mistaken in the belief that the government planned to prohibit women from all parts of the white lead trade.⁶¹

Interestingly, men working in the white lead trade did not make any representations to the Home Office on the subject of their female co-workers. However, resolutions were passed at both the 1894 and the 1895 Trades Union Congress supporting the clause that would give the Home Secretary power to restrict or prohibit women's work in dangerous trades. In his 1894 presidential address, Frank Delves said that he did not want to propose any reactionary measures nor advocate closing trades to women except those which 'lead to the brutalising and poisoning of them'.⁶² Hence, they had a strong case for accepting the Home Secretary's proposal. The following year, the congress voted for the declaration of more trades as dangerous and the implementation of special rules and regulations in them. But in the end it was the women in or affiliated with the trade union movement that displayed the most interest in the white lead proceedings and pressed for the extension of the government's power regarding women's work in dangerous trades.

'Apart from the expression of opinion by the organised women', Tuckwell of the WTUL wrote, 'there has been for some time a growing feeling – roused I think in great measure by the series of articles which appeared in the *Daily Chronicle*, entitled "Death in the Workshop" – that some special legislation should be directed in guarding the dangerous trade.'⁶³ She and other members of the WTUL believed that lead poisoning was a special women's problem. And anyone, she argued, who had visited white lead or pottery factories or studied the evidence uncovered by governmental committees could not doubt the necessity of safeguarding those industries. White lead mills, Tuckwell wrote, were commonly known as 'death traps'. After visiting several of them she could tell tales of women becoming paralysed or blind and death stories 'the details of which I forbear to give since they would sound sensational'.⁶⁴ Most of the women were young but 'it was impossible to guess from their appearances, since those engaged in the most dangerous processes were very haggard, with decaying teeth and blue lips'.⁶⁵ In the end, she concluded 'the worst feature of this trade was its effect on the next generation'.⁶⁶ If children did survive birth, she added, they were usually sick and puny.

Evelyn March-Phillipps, of the Women's Cooperative Guild, also revealed

---

⁶⁰ C. E. Troup minute, 16 June 1894, ibid.
⁶¹ Benjamin A. Whitelegge minute, 17 June 1896, HO 45/9856/B12393AC.
⁶² *Annual report of the TUC* (1894), 31.
⁶³ Gertrude Tuckwell, *Women's work and factory legislation: the amending act of 1895*, London 1895, 7.
⁶⁴ Ibid.
⁶⁵ Ibid. 8.
⁶⁶ Ibid.

the hazards of this trade to women in an article in the *Fortnightly Review* in 1895.[67] The White Lead Committee's report and conclusions had convinced her of the urgent need for state intervention. In a manner reminiscent of the *Daily Chronicle*, she wrote:

> The deaths directly attributable to lead poisoning, and in which the proportion of women to men is ten to one, though quite sufficiently numerous, tell only part of the havoc wrought in human life by this deadly trade. Numbers of girls, suspended by the doctor's orders, take temporary employment and die from the effects of lead after an interval, and many deaths, due to illnesses induced by lead poisoning, are not attributed to that cause in the certificate of death. Death is common enough within two or three days of an attack.[68]

She told, as an example, the story of a young woman who was 'a tall fine, young woman when she began to work at the lead mill, but soon altered, and got white and wan enough. Taken ill on a Monday, she lay for three days in a terrible condition, and died on the following Thursday, with her knees drawn up against her mouth, in strong convulsions'.[69] Not only, she wrote, were women 'particularly sensitive' to the disease but their labour had dire consequences for their children. March-Phillipps had met a woman who had lost five children during her seventeen years in the lead trade. She knew numerous other cases of children who had died at birth or in infancy or survived only to live a stunted life. Consequently she believed that the proposed clause would help women rather then legislate them out of jobs.

These women also vigorously supported state intervention because of ample evidence that the special rules were not being followed in the white lead and pottery trades. March-Phillipps's article publicised the fact that only a few vigilant factories had stringently implemented the special rules. In many places, she contended, insanitary conditions led to the spread of lead poisoning while many foremen and doctors were indifferent to the women's suffering. Many workers had complained to Emilia Dilke, president of the WTUL, about the insufficient protective gear which employers were legally required to provide for the workers. One lead-poisoned woman had worn her respirator and overalls but once they had worn out the company did not replace them. Regarding the contention that the workers were negligent, Dilke wrote,

> I would reply that on more than one occasion, I have found, on inquiry, either that the baths were so foul from use that additional danger was incurred by

---

[67] For more on the activities of this organisation see Jean Daffin and David Thomas, *Caring and sharing: the centenary history of the Co-operative Women's Guild*, Manchester 1983.
[68] Evelyn March-Phillipps, 'Factory legislation for women', *Fortnightly Review* lvii (1895), 741.
[69] Ibid.

washing in them, or that no towels were supplied, so that the bather's own clothes were the only means she had of drying her body after bathing.[70]

Meanwhile other members of the WTUL reported that in the pottery trade, in which women also suffered from lead poisoning, 'regulations as to the supply of efficient precautions tend to become a dead letter unless supported by constant and competent inspection'.[71] To make matters worse, unionism could not improve women's working conditions in such dangerous trades. It was non-existent in the white lead trade with no prospect of its appearance in the future. The WTUL had made inroads only in the pottery trade in 1893 and at this point the North Staffordshire branch had no more than around 200 members. In view of this situation Dilke argued that 'our best hope is in increased inspection and such legislation as, in the national interest, may check the employment of women in these, to them especially, dangerous trades'.[72] This followed their overall strategy of looking to government intervention in order to raise the standards of women's work because unionism, at present, was not powerful enough to achieve that end.

These, then, were the pro- and anti-protectionist views expressed by women on the subject of Asquith's proposed dangerous trades clause. The extension of the Home Secretary's power over women's work was yet another contentious point between them but there was more at stake here; a government official had never been empowered to abolish women's work in any trade simply by decree. Both sides sympathised with the plight of women working in dangerous trades and agreed on the need for more stringent implementation of the government's special rules but they differed over the issue of removing women from the trade. By deconstructing the official reports, the SPW and WIDC attempted to show that lead poisoning was not peculiar to women but resulted from the disregard of precautionary measures and other circumstantial factors. The WTUL, on the other hand, was convinced by sensational reports of death among women in the trade that women were peculiarly susceptible to the disease. Steeped in the dominant social ideal of protecting women, they endorsed the prohibition of women because it would mean the elimination of sickness and death.

Eliminating women from dangerous work in the white lead trade

Asquith's proposed dangerous trades clause elicited only minor discussion in parliament. Newspaper reports of death in white lead workshops and the findings of the White Lead Committee had apparently convinced the lawmakers that the government had to take more stringent steps to combat

---

[70] Lady Emilia Dilke, *The industrial position of women*, London 1895, 9.
[71] Ibid.
[72] Ibid.

the lead-poisoning problem among women. The few objections that were raised in the House of Commons centred on whether or not the Home Secretary should have such extensive discretionary powers. 'Under the power to prohibit the employment of certain classes of persons in certain employments', former Home Secretary Matthews said, 'the influence of a Trade Union might be brought to bear to prohibit the employment of women in certain classes of work, although there was no reason to think these employments were in themselves either dangerous or undesirable.'[73] He and C. B. Stuart-Wortley, a well-known advocate of women's rights, were the only men to oppose the dangerous trades clause out of suspicion of the unprecedented power it bestowed upon the executive. The clause, Stuart-Wortley told his peers,

> not only applied to the employment of women, and was directed, possibly, at restricting, if not prohibiting, their employment in certain industries; but the clause went further, because its object evidently was to withdraw this kind of prohibition from the cognisance and control of this House.[74]

He aptly perceived that this clause would effectively remove parliament from the process of creating dangerous trades regulations. Both men also felt that dangerous trades regulation had strayed from its original purpose. Matthews, who framed the 1891 clause allowing for the construction of special rules, had anticipated that it would cover dangerous conditions, such as excessive humidity in workshops or factories. This point was echoed by Stuart-Wortley who thought it was intended to enable the Home Secretary 'to bring up the regulation requirements as regards machinery and processes to the best and latest standard of efficiency and protection'.[75] Consequently, both mistrusted this radical proposal that moved well beyond issues of sanitary regulations and standards.

Henry J. Tennant, former member of the White Lead Committee, easily deflected their objections by quoting passages from the government's investigations. That committee had found horrible illness in the trade, he said, and had considered several alternatives: forbidding the production of white lead, altering the method of its production, imposing certain restrictions or excluding women from its manufacture. In view of the insidious and deadly nature of the poison, they made the drastic recommendation that women be excluded from all direct contact with lead. Moreover, the recent revelations of illness and death in the trade, Tennant added, certainly refuted the arguments 'that it was better that some risk should be run than a large number of women should be thrown out of the employment and turned out into the

---

[73] *Hansard*, 4th ser. xxxi. 186 (1 Mar. 1895).
[74] Ibid. xxxii. 1408 (22 Apr. 1895).
[75] Ibid. xxxi. 199 (1 Mar. 1895).

streets'.[76] The committee had estimated that no more than 600 women would be displaced due to that measure which, he thought, was not too drastic to safeguard the health and life of workers.

Once his proposal became part of the 1895 Factory and Workshop Act, Asquith consulted the chief inspector of factories about the expediency of carrying out the recommendation of the White Lead Committee. Oram solicited information and opinions from factory inspectors and, not surprisingly, they endorsed the removal of women from the white lead trade. Richard Johnson of Newcastle-upon-Tyne, for example, remarked that

> Judging from the conversations I had with the employers and work managers, I consider the time is ripe for the proposed change, the managers without exception would welcome it. They are heartily tired of the onerous duties imposed upon them in connection with women's employment, and say that the change will be an agreeable one to them, whatever, it may be to the employers and employees.[77]

Women, they said, tricked them when they tried to enforce the special rules and, in general, women were hard to control. He and other factory inspectors, like A. P. Vaughan, believed that displacement would be beneficial. Vaughan, who admitted that he had always held the strongest opinion that women should be prohibited because of their constitutional peculiarity, wrote to Oram that, 'even if the employment of men caused in the first instance a slight increase in the cost of production, this would probably be compensated by the increased amount of work which men could do'.[78] Finally, Edward Gould, the superintending factory inspector who had chaired the White Lead Committee, had read the factory inspectors' reports and wrote to Oram 'I find, as expected, that there is practically complete accord as to the desirability of carrying out the recommendation in question.'[79] Week after week, he added, fresh evidence substantiates the fact that lead is poisonous and 'I cannnot but think it worse than imprudent to allow the comparatively few persons of the sex which is especially liable to plumbism, to incur any longer the pernicious consequences of their employment in lead in its most dangerous form.'[80]

The proposed prohibition of women went into effect, as recommended, in June 1898. Women could no longer work in the white beds, rollers, washbecks, stoves or in packing lead, in other words, the most dangerous yet highest-paying parts of the trade. After that date men replaced women in those jobs, and they became ill in increasingly large numbers. Government statistics for lead works in Newcastle-upon-Tyne (*see* chapter 6) showed that

---

[76] Ibid. xxxii. 1411–12 (22 Apr. 1895).
[77] Report, Richard Johnson to Cooke-Taylor, 2–3.
[78] A. P. Vaughan to Oram, 15 Feb. 1896, HO 45/9856/B12393AC.
[79] Edward Gould to Oram, 25 Feb. 1896, ibid.
[80] Ibid.

between 1 December 1897 and May 1898, nineteen males and sixty-six females were reported ill, while betweem 1 June 1898 and 30 November 1898, when males first displaced females, according to government regulations the figures were eighty-two and twelve respectively. Statistics for in-patients at the Newcastle Infirmary showed an increase in male cases of lead poisoning from seven in 1897 to twenty-two in 1898, nineteen in 1899 and fourteen in 1900, while the number of females treated decreased from twelve for both 1897 and 1898 to one in 1899 and none in 1900.[81] These statistics showed that as the number of men working in the most dangerous portions of trade increased so did their incidence of lead poisoning, thereby undermining the theory that sexual difference was the decisive factor in the development of the disease.

Despite the foregoing statistics, proponents of eliminating women from the trade did not doubt that the government had taken the appropriate action. Confronted with statistics of increased illness among men, Oliver admitted that they might raise doubt as to the greater susceptibility of women to lead poisoning and that it might, in fact, be equal in the two sexes. Still, he concluded 'that in the main the symptoms are neither so severe in man, nor does the malady run so rapidly to a fatal termination as it does in women'.[82] He then referred to lead work and infant mortality. Dr Thomas Legge, the newly-appointed medical inspector of factories, concluded that the statistics 'do not lead to the conclusion that females are more susceptible to lead than males' but added 'the influence of lead on the child-bearing function is of immense importance'.[83] Even when medical theory, which justified the elimination of women from the trade was undermined, the government's actions could be vindicated on the basis of married women's work and infant mortality.

Nor did the government's statistics change public opinion. By 1898 fury over female labour in the white lead trade had subsided. Agitation and interest had climaxed in 1894 and 1895; it was old news now. The fact that male workers were suffering in the trade did not interest or alarm people for women's work in a second lead trade, the pottery trade, occupied centre stage in the dangerous trades arena until the First World War.

[81] These statistics were reprinted in Oliver, 'Lead and its compounds', 297–8.
[82] Ibid. 298.
[83] Dr Thomas Legge, 'Industrial lead poisoning', *Journal of Hygiene* i (1901), 103–4.

4

Dangerous Trades Regulations and the Pottery Trade

The Pottery towns of Stoke-on-Trent had been renowned for their beautiful china and earthenware since the late eighteenth century. Wedgewood and Dalton wares, for example, had long graced the dining rooms of people in England and Europe. In the decades before First World War, however, attention was not focused on the beautiful wares produced in the Potteries but on the lead poisoning that was decimating its working population. Workers were exposed to two dangers in the trade: pulmonary disease resulting from the inhalation of clay dust and flint as well as lead poisoning from the lead used in pottery glazes and the colours used for decoration. Blue lines on gums, colic, stomach disorders and paralysis were the symptoms of lead poisoning. In 1895 a little over 27,000 women were employed in this trade.[1] They worked as dippers and dippers' assistants who dipped the ware into leaded glazes or rubbed the glaze off the ware, as glost placers who placed the dipped ware into the stove to dry and as majolica paintresses who painted patterns on the wares. They also worked as scourers who brushed the dust from the pottery after its removal from the oven.

The trade was investigated as part of the new avenue of dangerous trades regulation and declared dangerous on 24 December 1892. Special rules were drafted in February of 1893, and this was followed by the appointment of a special departmental committee to investigate the conditions of labour in the Potteries and their effect upon the health of the workers. The committee recommended several possible remedies for the lead-poisoning problem including the enactment of special rules and the prohibition of women from the trade.[2] In the end, the government chose to implement special rules. They first came into effect in 1894, were amended in 1898, amended again in 1903 after a lengthy arbitration, and once more in 1913. According to them, women were subjected to medical inspection and possible suspension, prohibited from working in the preparation of lead glazes and certain 'heavy work' or work involving strain without a certificate of permission.[3] Although

---

[1] Harrison, *Not only the 'dangerous trades'*, 60.
[2] *Report of the departmental committee on the conditions of labour in the Potteries, the injurious effects upon the health of the workers, and proposed remedies*, PP 1893–4, [c.7240] xvii, 47.
[3] The various measures are outlined in Dame Adelaide Anderson, *Women in the factory: an administrative adventure, 1893–1921*, London 1922, appendix i.

less drastic than the white lead measures, they none the less constituted significant measures to safeguard women and their potential offspring.

This chapter will examine the creation of those regulations and suggest that both similarities and differences between the two lead trades prompted and influenced their creation. First, and most significantly, medical and literary opinion persisted in the presentation of lead poisoning as a 'woman's problem'. The *Daily Chronicle* led the press in the creation of sensational stories about female labour in the trade while medical men continued to argue that sexual difference was the key to understanding the development of the disease. As a result, the government's investigations primarily focused upon women and, especially, the impact of this labour upon their unborn children. Their elimination from the most dangerous, and highest-paying, portions of the trade was discussed throughout the protracted proceedings. The final outcome of this episode in government intervention, I will argue, was shaped by the diverse ways in which lead was used in the trade, the fact that women constituted a larger percentage of the workforce than in the white lead trade, and a vocal lobby of manufacturers who wanted to continue to employ them. In this more complex situation, the English government compromised its desire to protect endangered married women and their potential offspring with a desire not to unduly or adversely affect business interests.

The 'discovery' of the lead-poisoning problem

Progressive medical men were writing about the dangerous nature of the pottery trade in the mid-nineteenth century.[4] Dr John Simon, medical officer to the privy council, directed the systematic investigation of a variety of industrial diseases. Dr Edward Greenhow carried out an investigation of thirty-three industrial towns for the privy council, including the Potteries in Stoke-on-Trent, in 1860 and 1861. He concluded that potters had a far higher mortality rate due to phthisis (consumption) and other respiratory and pulmonary illness, especially bronchitis, than the general population of the area. While noting that other factors might contribute to the incidence of those ailments, he ultimately related it to the nature of the trade.[5] Greenhow's opinion was seconded by Dr John T. Arlidge who immediately noticed the potters' poor physical condition when he became physician to the North Staffordshire Infirmary, Stoke-on-Trent, in 1862.[6] 'The potters, as a class, both men and women', he said,

---

[4] For more on the subject of industrial illnesses see Anthony Wohl, *Endangered lives: public health in Victorian Britain*, London 1983, ch. x.
[5] Ibid. 261–2.
[6] For more on Arlidge's activity see Clare Holdsworth, 'Dr John Thomas Arlidge and

represent a degenerated population both physically and morally. They are as a rule stunted in growth, ill-shaped and ill-formed in the chest. They are certainly short-lived; they are phlegmatic and bloodless ... of all diseases they are especially prone to chest disease, to pneumonia, phthisis, bronchitis, and asthma. One form would appear peculiar to them, that which is known as potter's asthma or potter's consumption.[7]

Arlidge, Greenhow and Simon, in particular, believed that the high incidence of illness among pottery workers was serious enough to warrant government intervention. The government responded to their findings with modest provisions in the 1864 Factory Act. This legislation attempted to tackle insanitary conditions by requiring the erection of ventilation systems in trades where manufacturing produced dust, gas or other impurities in the air. In addition it stipulated that women and children could not eat their meals in the dipping houses or drying rooms where lead was used in the glazing process.[8] The government's response was modest because of the prevalent fatalistic attitude towards industrial illness – an attitude shared by workers, employers and the government. It was commonly observed that workers in different trades developed certain illnesses but that was no cause for alarm. At least, that is, until the autumn of 1892.

William Owen and Thomas Edwards, members of the Amalgamated Potters' Society and United Firemens', Dippers', and Placers' Association, drew attention to the unhealthy nature of the trade when they testified before the 1892 Royal Commission on Labour. In November 1892 they discussed the two hazards of the trade: pulmonary diseases and lead poisoning. When asked about the effect of using fans to remove dust from the workshops, Owen launched into a scathing indictment of the bad working atmosphere. He remarked:

> The process has only been in operation perhaps about eight or nine years. There have been married women following in it [the trade] whose lives have been shortened and whose children have died through it after they had been born. A doctor in the town in which I live said that a woman who followed one of the dusty employments could not have a child live because its breathing organs were affected by it.[9]

Edwards emphasised a previously overlooked hazard of the trade, lead poisoning resulting from the use of lead in the pottery glazes. He recom-

---

Victorian occupational medicine', *Medical History* xlii (1998), 458–75, and E. Posner, 'John Thomas Arlidge and the Potteries', *British Journal of Industrial Medicine* xxx (1973), 266–70.
[7] Arlidge cited in Posner, 'John Thomas Arlidge', 267.
[8] For more on these early developments see Hutchins and Harrison, *A history of factory legislation*, 151–4.
[9] *Royal Commission on Labour: third report, digest of evidence, group c*, PP 1893–4, [c.6894–xii] xxxiv. 419.

mended that the government restrict young persons from working in the dipping houses, where the glaze was found, and the use of lead in glazes.

The testimony of Owen and Edwards captured the attention of the *Daily Chronicle*. On 14 and 24 November it published two highly sensational articles entitled 'Death in the workshop: through the Potteries: the dust death' and 'Death in the workshop: through the Potteries: dust and poison and the remedies'. 'Dismal by day, thrice dismal by the night, are the Potteries', the special correspondent wrote after his visit to Stoke-on-Trent. 'It is a dismal region', he continued,

> A dense tight-clinging pall of smoke shuts out the sunlight and the sky. In damp and rainy weather the moisture mixes with the coally atmosphere and smears itself over everything. A long course of this treatment means the Potteries are ingrained with black like an iron or coal workers hand. The streets and pavements in this November weather are coated with black slime, and so too are the ragged gaps in the towns which serve as open spaces but which are too nasty for even children to play in.[10]

Thus, he described the towns which were world-renowned for their exquisite pottery. ' "Dust to Dust" we accept as the voice of the unalterable', he wrote, 'when life is fled. But dust to dust as a daily doom, to be buried alive by slow suffocation, is more horrible than death. This is the deadly depression which strikes the visitor to the Potteries.'[11]

The union men's testimony and the newspaper articles stimulated further activity. The trade was officially designated dangerous and special rules were drafted early in 1893.[12] Dr Arlidge took the opportunity to acquaint the government with his medical knowledge. In addition, Clara Collett, an assistant commissioner for the Royal Commission on Labour, was instructed to investigate the conditions of women's work in the trade while the local factory inspector, Dawkins Cramp, was asked to provide a general report on the trade. Finally, pottery manufacturers entered the fray. These developments signalled the beginning of tremendous interest in the Potteries and the end of the prevalent fatalistic attitude towards its specific industrial diseases.

Dr Arlidge sent a copy of his pamphlet, *The pottery manufacture in its sanitary aspects*, to the Home Office in April 1893.[13] It included statistics for 463 male and 337 female out-patients treated at the North Staffordshire Infirmary.[14] His records of the various types of illness among these male and

---

[10] *DC*, 14 Nov. 1892.
[11] Ibid. 24 Nov. 1892.
[12] The rules, drafted on 28 Feb. 1893, were enclosed in Oram to Lushington, HO 45/9851/B12393E.
[13] Dr John Arlidge, *The pottery manufacture in its sanitary aspects*, Hanley 1892. This pamphlet was sent with a letter, 23 Apr. 1893, HO 45/9851/B12393E.
[14] Marguerite W. Dupree has noted that the Infirmary was financed through donations from individuals and subscriptions from firms, such as pottery firms. The latter's subscription came from deductions from workers' wages and, if a worker became ill and received a recom-

female pottery workers (*see* table 1) show that the highest percentage of males were treated for bronchitis, females for stomach disorders, while phthisis was second on the list of illnesses for both sexes. Eight per cent of the male and 5.06 per cent of the female patients were suffering from lead poisoning. Arlidge acknowledged constitutional peculiarity as a factor in the development of lead poisoning but concluded:

> nevertheless, in the majority of cases, the difference can by explained on more obvious grounds, existing in carelessness at work, indifference to cleanliness of the person and clothing, in reckless eating and drinking in the place of work, and wearing clothes begrimmed with glaze, whereby the poison is carried to the homes. Lastly, ill-constructed, dirty, confined shops and want of free ventilation contribute an important factor.[15]

Despite the higher incidence of lead poisoning among men and his emphasis on circumstantial factors contributing to the development of the disease Arlidge wrote:

> Considering that women and children are more readily affected by poisons, a good reason is found for excluding them, as far as possible, from the dipping-house and dipping-tub. The dress of women and their long hair furnish a media for the accumulation and carrying away of the poisonous dust, and far too commonly they are unwilling to lessen their proclivity by suitable coverings.[16]

It is interesting that Arlidge had not previously linked sexual difference to the development of lead poisoning. In fact, he made a conscious effort to outline the various circumstantial factors connected to the development of the disease.

The reports of Collett and Cramp likewise linked circumstantial factors to the development of the disease and spoke of the higher incidence of illness among male workers. 'The men', Collett reported,

> struck me as looking much more unhealthy than the women. In some cases the girls in the glazing department look very white and ill; the 'towers' look very rough and were said to be hard drinkers; with these exceptions it would be difficult to find so many good looking girls as are to be seen in any one of these factories.[17]

---

mendation from the firm, they would be treated at the Infirmary. For more on the North Staffordshire Infirmary see Marguerite W. Dupree, *Family structure in the Staffordshire Potteries, 1840–1880*, Oxford 1995, 293–7.
[15] Arlidge, *The pottery manufacture*, 15.
[16] Ibid. 16.
[17] *Royal Commission on Labour: report of the lady assistant commissioners*, 62.

Table 1
Illness among pottery workers, by sex

Illness	Males (%)	Females (%)
Bronchitis	36.57	7.14
Phthisis	20.90	16.96
Rheumatic affections	7.79	4.46
Stomach disorders	8.44	19.64
Plumbism	8.00	5.06
Cerebro-spinal diseases	4.32	2.97
Cardiac diseases	2.81	2.08
Epilepsy	1.73	4.46

Source: Arlidge, Pottery manufacture, 9.

Cramp reviewed the statistics of illness from the North Staffordshire Infirmary and noted the slight preponderance of men suffering from lead poisoning. He continued:

> Of course some persons are more susceptible than others to the influence of lead poisoning, and whilst there are some instances of men working at dipping for 30 or 40 years without injury, others are maimed or invalided for life, or even die, in six months.[18]

He concluded that personal cleanliness and work habits were critical factors related to the incidence of illness.

Pottery manufacturers immediately felt harassed by the press, government and medical men and denounced state intervention in the trade. They challenged the process of declaring a trade 'dangerous' in their trade journal, the *Pottery Gazette*, and in meetings of the North Staffordshire Chamber of Commerce and Staffordshire Potteries Manufacturers Association.[19] The press was implicated in this unjust action because it had publicised the workmen's prejudicial statements. 'The highly coloured, sensational articles of the *Daily Chronicle*', a writer for the *Pottery Gazette* claimed,

> have not been without their effect on the more credulous or ignorant portion of the public, and these, together with the statements before the Royal Commission, have resulted in the representatives of the workmen to frame special rules for the use of earthenware and china manufacturers, with a view to their adoption by the Home Office.[20]

---

[18] Report, Dawkins Cramps to Oram, 19 Nov. 1891, HO 45/9851/B12393E, 3.
[19] Arthur Llewellyn, secretary of the North Staffordshire Chamber of Commerce, sent their first objection to the designation of their trade as 'dangerous' to the Home Office in a letter, 20 Jan, 1893, ibid.
[20] PG, 2 Jan. 1893. This point was also made directly to Asquith by a deputation from the

A leading manufacturer also criticised the Home Secretary's actions at a Chamber of Commerce meeting: 'Under the Factory Act, 1891, he [the Home Secretary] had unlimited powers, and he used them in a way that seemed best to him by appointing gentlemen to gather information for him without any reference at all to the manufacturers.'[21] Both remarks reflect the manufacturers' contention that they had been excluded from this dubious process. Lastly, they attacked Dr Arlidge's work because of its emphasis upon links between the insanitary conditions of the workplace and illness. 'The medical men of the present day', the editor of the *Pottery Gazette* wrote 'appear to live on microbes; their existence seems to be a state of feverish unrest consequent on the ceaseless activity they displayed in the pursuit of their prey in all directions; this is somewhat bewildering to the ordinary business mind.'[22] These ordinary business minds argued that their trade was no more unhealthy than other trades, echoing the earlier naturalistic attitude towards industrial hazards, and that they had been unfairly singled out for government intervention. As far as they were concerned, careless workers were to blame for work-related illnesses.

The Home Office dismissed the manufacturers' argument that their trade was not dangerous and, after a year of protracted comments from them and the male pottery workers, the first set of special rules were enacted in 1894. Employers agreed to provide overalls and head coverings for women (they had won their case against such provisions for men), fans to remove dust from the scouring processes, soap, water and brushes for the employees to wash with, and brooms and other items required to clean the workshops. Workers, for their part, were required to wash before meals and clean their portions of the workshop while women were required to wear the protective gear provided for them. This proved to be the first of several confrontations between the government and pottery manufacturers.

Reproductive danger in the potteries

By the time these special rules had been enacted, government and public attention had been seized by the scandal of women's work in the white lead trade which eclipsed concern about the general conditions of work in the Potteries until its resolution in 1898. There were two notable developments in the white lead case that influenced further proceedings in the pottery trade: the appearance of Dr Oliver and the construction of lead poisoning as a

---

North Staffordshire Chamber of Commerce, the Staffordshire Potteries' Manufacturers' Association and the Plain, Decorative, Encaustic and Earthenware Tile Manufacturers' Association on 14 Mar. 1894. The proceedings were published in *The Times* and the SS on 15 Mar. 1894.

[21] The proceedings of the North Staffordshire Chamber of Commerce meeting, 21 Mar. 1894, were reprinted in the SS on 24 Mar. 1894.

[22] Ibid.

'woman's problem'. Oliver was hailed as the medical expert on lead poisoning and with his claim that women were 'naturally' more prone to lead poisoning, the incidence of lead poisoning was, for the first time, linked to sexual difference. Suddenly, biological differences between men and women provided the explanation for this illness, and established, contrary to earlier opinions, that it was a problem peculiar to women. The popularisation of this theory and fear about the impact of lead work on unborn children led to the radical resolution of banning women from dangerous portions of the trade and set an important precedent. The impact of these developments can be seen by examining two pottery investigations conducted by the government.

General evaluations of the trade were replaced by the evaluation of the 'greater risk' female workers. Moreover, the government's investigation was narrowed even further to revolve around the critical issue of married women's work and childbearing. For instance, the 1893 departmental committee appointed to investigate the general hazards of pottery work devoted a tremendous amount of attention to that particular subject and offered two significant conclusions.[23] First, they noted that 'the higher rate of wages obtainable by dipping house labour acts as an inducement to women and young people to engage in it, also to continue to work even when aware of its ill effect upon their health'.[24] Second, and most important, they pronounced 'It is sufficiently established that lead may provoke miscarriages when a woman is pregnant, and likewise operates prejudicially upon the unborn child, if it survives its birth.'[25] To counter those problems they suggested the medical examination of all workers who came in contact with lead and the removal of married women employed in processes in which lead was used. They were persuaded, no doubt, by such testimony as that from a MOH who said 'Abortion and premature births are not uncommon in women poisoned by lead and dust. . . . The vigour or decline of a great portion of the present, rising, and future generations depends upon the inquiry leading to satisfactory legislation on the subject.'[26] He then recommended further state intervention to deal with this disturbing problem.

The factory department also sent Factory Inspectors Lucy Deane and Mary Paterson to North Staffordshire in 1897 to assess the effects of the special rules on illness and, more particularly, the relationship between married women's work and infant mortality.[27] Regarding the special rules, Deane

[23] This committee included Factory Inspectors May, Cramp and Walmsley as well as Dr Arlidge and Dr Spanton, a surgeon at the North Staffordshire Infirmary.
[24] *Report of the departmental committee on the conditions of labour in the Potteries*, 47.
[25] Ibid. 48.
[26] Ibid. 60.
[27] Male hegemony in the Factory Department had only ended in 1893 when May Abraham and Mary Paterson were appointed to this position. Three additional women were added to the department over the next two years. They were appointed with the stipulation that they would travel around the country undertaking special investigations relevant to women workers. By 1897 the Potteries had become their 'unofficial' domain and they conscien-

commented that they 'were a dead letter and owing to wording not enforceable and if enforced no good'.[28] After visiting several firms and interviewing numerous victims of lead poisoning, she and Paterson concluded that government intervention thus far had failed to improve working conditions in the Potteries. Most significantly, they devoted a considerable amount of time to recording illness among married women. Out of seventy-seven married women reported ill between 1896 and March 1897, they reported, '15 have been childless and have had no miscarriages; 8 have had 21 still-born children; 35 have had 90 miscarriages, and of these 15 have had no child born; 36 have had 101 living children, of whom 61 are still alive, the great majority of the 40 who are dead have succumbed to convulsions in infancy'.[29] Those statistics indicated that infant mortality was indeed a serious problem yet Deane and Paterson recommended the medical examination and suspension, not prohibition, of women in the trade.

These female government officials were keenly aware of the implications of their investigation; their findings could be used to make lead poisoning in the pottery trade (as in the white lead trade) a peculiar 'woman's problem'. According to Mary Drake McFeeley, they collected statistics on the incidence of illness among women and men in order 'to blunt the risk of making lead poisoning a "women's problem" that could be solved by eliminating women from the workplace'.[30] Likewise, they did not want to take the drastic step of simply eliminating women from the trade because they appreciated the obscured but seminal point of the discussion of married women's work: economic necessity. They astutely observed that

> The time when the need for money is the most urgent, when, for some weeks in the near future, the earning power is certain to be suspended, is not the time when a women's judgment can enable her to balance justly her future loss against the immediate gain, which appears to her so important. The advice of the doctor is disregarded, and nothing at present stands between women and the suicidal course to which they are drawn.[31]

Sensitive to the plight of working women, they believed that simply eliminating them from the trade would not solve the problem. They would, in the future, repeatedly urge the government to explore and exercise other options.

The discussion of this highly charged subject moved from restricted governmental circles to the public domain in 1898. Once again the press was

---

tiously applied themselves to the task of investigating the lead poisoning problem. For more on the subject see Helen Jones, 'Women health workers: the case of the first women factory inspectors in Britain', *Social History of Medicine* i (1988), 165–81, and Mary Drake McFeely, *Lady inspectors: the campaign for a better workplace, 1893–1921*, Athens, Ga. 1991, ch. ix.

[28] McFeely, *Lady inspectors*, 65.

[29] *Report of the chief inspector of factories and workshops for 1897*, PP 1898, [c.8965] xiv. 53.

[30] Deane made this point in her diary: McFeely, *Lady inspectors*, 66.

[31] *Report of the chief inspector of factories for 1897*, 54.

instrumental in portraying the particular dangers of this work for female workers. Readers of papers such as the *Daily Chronicle*, the *Star*, the *Pall Mall Gazette* and the *Penny Illustrated Paper* encountered stories of afflictions ranging from blindness to infant slaughter. A provincial paper, the *Bradford Observer*, was struck by the numerous stories 'of young women of fine natural physique disfigured and tortured through a shortened life by loss of sight, speech, strength, mental facilities – worse than that, giving birth to children who are old and decrepit at birth, or breaking down in the office of maternity'.[32] The *Daily Chronicle* distinguished itself with a series entitled 'Lead in the home' which included 'Whole families desolated: a story of mothers and children' or 'Infanticide in the Potteries: weeping for her children because they are not'. As will be seen in the next chapter, they hoped that graphic representations such as these would prompt the government to enact more stringent measures for the trade.

The press also publicised and supported a campaign, led by MPs Charles Dilke and John Burns, and the WTUL, for the appointment of a resident female factory inspector for the Potteries in Staffordshire.[33] In petitions and deputations to the Home Office, this lobby also contended that lead poisoning was a special problem for women. Stationing a specialised factory inspector would, they argued, contribute to its resolution because she could ensure the strict enforcement of the government's special rules. They demanded a woman inspector because of the high concentration of women in the pottery works in that area: at the time, 20,000 of the women working in the trade were employed in that geographical area. This request stemmed from their high regard for the female inspectors' abilities as well as the belief that such an appointment would facilitate communication between women workers and factory inspectors. One petition emphasised that women felt uncomfortable talking about personal information with a male inspector with the result that many cases of lead poisoning would not be reported.[34]

The men in the Home Office bluntly rejected this request. Besides disqualifying the women on the grounds that they lacked medical knowledge, Chief Inspector of Factories Whitelegge felt that such an appointment would involve 'administrative difficulties'.[35] He anticipated conflict between the local district supervisor and the woman inspector who was not responsible to him but to the principal lady inspector. C. E. Troup, under-secretary at the Home Office, not only agreed with Whitelegge but had earlier made the

---

[32] *DC*, 30 June 1898.
[33] For a more in-depth analysis of their campaign see Peter Bartrip, ' "Petticoat pestering": the Women's Trade Union League and lead poisoning in the Staffordshire Potteries, 1890–1914', *Historical Studies in Industrial Relations* ii (1996), 3–26.
[34] Sarah Bennett to Home Secretary Matthew Ridley (enclosing petition), 17 Apr. 1898, HO 45/9933/B26610. Gertrude Tuckwell, the WTUL, the duke and duchess of Sutherland and workers were among the 400 signatories.
[35] Whitelegge minute, 22 Apr. 1898, HO 45/9933/B26610.

disparaging remark, 'I suppose this may be taken as indicating the direction in which the friends of the Lady Inspectors' Department propose now to agitate for extension of their powers.'[36] His remark about extending the scope of the women factory inspector's work reflects the resentment which some of the men in the Home Office felt towards them.[37]

This curt dismissal did not put an end to the matter; it had become a labour issue. The Home Office was flooded with further requests for a resident women inspector from such groups as the WTUL, the Staffordshire Trades and Labour Council and the Midland Federation of Trades Councils.[38] Most significantly, MPs Dilke, Burns and Henry J. Tennant, put further pressure on the Conservative Home Secretary, Matthew Ridley, at a highly publicised deputation on 19 May. This meeting garnered tremendous attention because six sick male and female pottery workers travelled to London to personally relate their tale of suffering to him. Ridley's refusal to meet them caused quite a stir, but the deputation went on as planned with its leaders primarily pressing for the appointment of a resident women factory inspector for the district. Dilke said that

> although the official returns showed that men were the greatest sufferers from this dire disease, local inquiries went to show that the contrary was the case. The increasing number of cases of blindness among women and girls employed in the Potteries had given rise to the feeling that a women inspector for the Potteries should be appointed.[39]

His point was reinforced by Mrs Walsh, a member of the Hanley branch of the WTUL, and Mrs Greatorex, representing non-union pottery workers, who testified as to the ill-effects of pottery labour for women. Other members of the deputation put forward their secondary demands which included the medical examination of workers and the extension of the 1897 Workmen's Compensation Act to lead poisoning victims.[40]

Confronted with this groundswell of pressure to safeguard pottery women, Ridley made the widely unpopular decision not to appoint a resident female factory inspector.[41] He sought to appease those who supported that course of action with the announcement that he had already taken two significant steps in the battle against lead poisoning. First, he had drafted a new set of

---

[36] Troup minute, 22 Apr. 1898, ibid.
[37] McFeely, *Lady inspectors*, 27–33, 75.
[38] Ridley received letters from Mona Wilson, secretary of the WTUL, 27 Apr. 1898, Isaac Harvey, secretary of the North Staffordshire Trades and Labour Council, 25 Apr. 1898, and S. Middleton, Midland Federation of Trades Councils, 17 May 1898, HO 45/9933/B26610.
[39] *DC*, 20 May 1898.
[40] The 1897 Workmen's Compensation Act has been analysed in Bartrip and Burman, *The wounded soldiers of industry*, 190–221.
[41] The government finally relented in 1902 and stationed Hilda Martindale, who had joined the Factory Department the previous year, in the Potteries: McFeely, *Lady inspectors*, 74–5.

special rules that included the monthly medical examination of women. Beginning on 1 August 1898, certifying surgeons would examine them and if they exhibited signs of lead poisoning they would be suspended, without pay, from work. In addition, he had appointed Dr Oliver and T. E. Thorpe to investigate other possible preventive measures for the lead-poisoning problem, including the viability of lessening lead in glazes or the introduction of a leadless glaze.[42]

After an extensive examination of potteries in Britain and the continent, Oliver and Thorpe presented their report to the government in 1899. It contained several important recommendations including specific ones regarding female and male workers. They discussed, once again, the greater susceptibility of women to lead poisoning and stated that

> The medical history of the pottery industry leads us, in the interests of women themselves, and to avoid the disasters which plumbism entails upon the progeny through the mother, to urge that women no longer should be employed in the dipping-house where the lead is used, and as ware-cleaners after dippers in lead glaze.[43]

Once again, Oliver viewed the issue from the perspective of reproduction and his idea of a preventive measure was the drastic one of eliminating women from certain work in this trade. Interestingly, it was also recommended that adult male dippers, dippers' assistants, ware cleaners and glost placers undergo systematic medical examination and, if they were found to be ill, be suspended from work.

In contrast to the white lead case, there was significant opposition to Oliver and Thorpe's proposal to ban women who came into contact with lead. Arthur Llewellyn, secretary of the Allied Manufacturers' Association, sent a statement from that group to Ridley claiming that such a regulation would deprive a considerable number of people of a means of livelihood

> especially suited to their strength and capacity, and at the same time would increase the wages cost in these departments, [it] is not called for if the use of 'unfritted' lead compounds in glazes be prohibited, and the monthly medical examination already in force is stringently carried out. The evidence collected in the Report, as to the experience gained at Continental factories, shows that, where the work is carried on under conditions suggested by us, women and young people are employed in these very occupations in considerable numbers, and enjoy immunity from 'plumbism'.[44]

---

[42] Dr Oliver and T. E. Thorpe received these instructions by letter, 28 Apr. 1898, HO 45/10117/B12393P.
[43] *Report on the employment of compounds of lead in the manufacture of pottery, their influence upon the health of the workpeople, with suggestions as to the means which might be adopted to counter their evil influence*, PP 1899, [c.9207] xii. 287.
[44] Statement from the Joint Committee of Various Manufacturing Associations to Ridley, 26 Apr. 1899, HO 45/9851/B12393E, 4.

This response was not surprising since manufacturers had a vested economic interest in employing women whose wages were less than those of their male counterparts. Moreover, the Staffordshire Potteries Manufacturers' Association and the North Staffordshire Chamber of Commerce had passed resolutions in 1894 and 1895 opposing the extension of the Home Secretary's power over dangerous trades in anticipation of its application to their trade.[45] One pottery manufacturer felt that allowing him to limit or prohibit the work of certain persons in dangerous trades was 'a dangerous departure in two ways; it not only restricted the employment of operatives more than had been previously done, but it conferred what he considered a very dangerous power upon the Home Secretary, whose position might depend on Parliamentary votes'.[46]

The two sides of the divided women's movement were unusually united in their criticism of the proposal regarding women. Boucherett read the report and called its statistics, reflecting a higher rate of lead poisoning among women, 'unfair and misleading'. The new rule mandating the medical examination of women would naturally raise, she argued, 'the number of cases among women; but the men being exempt from examination would show no increase of cases'.[47] The SPW sent a memorial to Ridley asking him to consider the adoption of precautions taken in foreign countries since Oliver and Thorpe's report illustrated that they had effectively eliminated lead poisoning among women and children.[48] Even Tuckwell of the WTUL, who had supported the ban on women in the white lead works, criticised this one in an article entitled, 'Commercial manslaughter'. The solution to the problem was not the prohibition of women but the prohibition of raw lead in glazes. She blamed the persistence of the lead poisoning problem on 'the extraordinary slackness and indifference of the state'.[49] And, she argued, it should force manufacturers to use leadless glazes in the production of their wares.[50]

Most significantly, virtually all the important figures in the Home Office failed to endorse Oliver and Thorpe's proposed course of action. Kenhelm Digby, permanent secretary at the Home Office, commented that 'they seem to me to have entirely gone beyond their province, and instead of containing

---

[45] See, for example, accounts of those meetings published in the SS, 18, 21 July 1894; 20 Mar., 23 Apr. 1895.

[46] The proceedings of the North Staffordshire Chamber of Commerce meeting, 21 July 1894, were published in the SS, 24 July 1894.

[47] Jessie Boucherett wrote an article on Oliver and Thorpe's report in ER xxx (1899), 100.

[48] Memorial, SPW to Ridley (on Oliver and Thorpe's report), reprinted ibid. 204–5.

[49] Gertrude Tuckwell, 'Commercial manslaughter', Nineteenth Century (Aug. 1898), 256.

[50] Tuckwell and Canon Charles Gore of the Christian Social Union had begun the Leadless Glaze Campaign, as it was called, in 1897. Over the next decade or so, they publicised the evils of lead, lobbied the government to ban the use of raw lead in manufacturing and organised Leadless Glaze Exhibitions displaying the 'leadless' wares for sale by 'enlightened' manufacturers. For more information about this campaign see the Christian Social Union's paper, Commonwealth, and Gertrude Tuckwell, Constance Smith: a short memoir, London 1931.

themselves to chemical and biological questions which they understand, to have endeavored to embrace the whole industrial and economic problems, without hearing the case for the employers and workers'.[51] Moreover, he wrote, this recommendation does not take into account the anticipated good effects of the monthly medical inspection of women. This report was sent to Chief Inspector Whitelegge who had seen the statistics of illness and death among the men who had replaced women in the white lead trade. Consequently he no longer believed in Oliver's theory of the greater susceptibility of women to lead poisoning. This influenced his reading of the report and led him to comment that

> the danger (naturally) varies much in the different processes, and the difference between the two sexes revealed lead poisoning is largely due to dissimilarity of employment. As it stands the report is likely to be regarded as more conclusion on the vexed question of female employment than it really is.[52]

Moreover, he maintained 'The experience of certain Continental factories mentioned in the Report seems to give hope that with due precautions comparative safety may be attained even when persons of both sexes are employed.'[53] Only Medical Inspector of Factories Legge agreed with the medical experts in this instance.

Heeding this advice, Ridley chose not to ban women from the lead-based parts of the pottery trade. However, the immense negative publicity surrounding their work affected them in other adverse ways. First, some pottery manufacturers took the initiative and began to replace females with male workers shortly after the *Daily Chronicle* began its publication of stories of infanticide. As McFeely has written, 'When reports of stillborn or sickly babies, poisoned in the wombs of women pottery workers, aroused public sentiment, the manufacturers, to avoid the label "baby-killers" would often replace women workers with men.'[54] Moreover, women who feared suspension from their jobs colluded with their employers to conceal their illness. These unofficial developments were certainly not the desired effect of protective labour legislation.

Arbitration: equal treatment for men and women?

The previous proceedings reveal that while the dangers of women's work and infant mortality became the primary focus of the discussion of this dangerous trade, there was a recognition that something should be done to protect its working men. To this end, Ridley adopted the recommendation of Oliver and

---

51 Kenhelm Digby minute, 21 Feb. 1899, HO 45/10117/B12393P.
52 Whitelegge minute, 2 Mar. 1899, HO 45/9851/B12393E.
53 Ibid.
54 McFeely, *Lady inspectors*, 66.

Thorpe and included the medical inspection of males who worked in the lead processes in his new set of proposed rules in 1899.[55] However, this was not presented as a *fait accompli* as it had been in the case of women. The proposed special rule was submitted to employers and working men and when the latter objected to it, unless they were to receive monetary compensation during suspension, the matter became one of several discussed at an arbitration which took place in November 1901 with an award issued in 1903.[56] The testimony pertinent to the medical inspection of men provides an insight into contemporary male opinions about gender and the concept of dangerous work.

First, virtually all the witnesses extolled the virtues of the medical examination and surveillance of women. As one long-time factory inspector in the Potteries testified, 'it had been of use in weeding out unsuitable and susceptible individuals. This is shown by the number of suspensions and warnings by the Certifying Surgeon'.[57] Dr Alcock, certifying surgeon for three and a half years in the pottery district of Burslem, felt that the periodic examination of women (and children) had successfully eliminated the feeble workers before serious symptoms could arise. Dr Legge likewise praised this policy for dealing with women workers and made some interesting comments on the vexed issue of women's work in the pottery trade. He noted that although recent statistics of illness among men who replaced women in the white lead trade had disproved the theory that women were more susceptible to lead poisoning than men, special precautions had to be taken because of lead's effect on women's reproduction.[58]

The medical examination and possible suspension of men was another matter entirely; witnesses urged modification in the procedure because of ideas about sexual difference and the persistent idea of a family wage. Dr Legge, for example, believed that men needed less frequent examination because they were attacked proportionally less than women. Moreover, anaemia, the chief cause of suspension among women, was found less frequently in them. He suggested an examination every three months but 'in view of the serious consequences which might result from the permanent

---

[55] Home Office officials had supported this proposal to extend medical inspection to men who worked in the leaded processes. Whitelegge notified Ridley that he had raised the issue with the four certifying surgeons for the pottery district and all but one agreed that it was a good idea: note, Whitelegge to Ridley, 7 Feb. 1900, HO 45/10118/B12393P.
[56] Their objections were raised in the Operatives Petition, 2 Apr. 1899, HO 45/9851/B12393E. Lord James of Hereford served as the umpire, Mr Cripps represented the Home Office, Arthur Llewellyn appeared on behalf of the manufacturers, and Mr Colefax presented the workers' case which was prepared by Mona Wilson, secretary of the WTUL. For more information on this arbitration see Bartrip, ' "Petticoat pestering" ', 20–2.
[57] Transcript of arbitration, proof of witnesses, 7–12 Nov. 1901, HO 45/10120/B12393P, 58.
[58] Ibid. 6. In the midst of the arbitration Legge published an article full of statistics on lead poisoning and married women's work: 'Industrial lead poisoning', 96–110.

suspension of men' he recommended limiting the power of the certifying surgeon to order the permanent suspension of males under thirty-five years of age from work.[59] Since, Dr Alcock contended, men 'have shown a comparative insusceptibility to poisoning', they should be examined every three months.[60] He thought that suspension from work should apply to men as well as women but that a variable standard of judgement should be used; each case should be judged according to the value of the worker to the family. Stressing the hardship male suspension would bring to the population in this area, Dr Arlidge argued 'There are men who have partial wrist drop, and seem to suffer in no other way, and are quite able to do their work as dippers or glost placers. To suspend such men would mean deprivation of food to themselves and [their] families.'[61]

Working men expressed similar concerns about state protection. William Ephraim Milner, a dipper representing the non-union men of Longton, testified that dippers and placers did not think that medical inspection was necessary for them. Indeed, they feared the results. He and Lord James, the umpire of the arbitration, had a very revealing interchange:

> Milner: We think if we showed the least trace of lead he (the Certifying Surgeon) would stop us.
> Umpire: Supposing you were not fit to work, why should not you be stopped?
> Milner: The question then is, who is going to keep a man and his family?
> Umpire: Perhaps the woman would not have a husband if he were not strong?
> Milner: No, but when they are bad they generally go to the club – there are clubs around here – if he is bad for a week, and when he is well he starts work again.
> Umpire: Even if you are going to your death?
> Milner: Certainly, we take our risks on ourself; we know the risk when we go working.[62]

The key issue here was the loss of wages that would result from a man's suspension from work. Thomas Edwards, who had previously appeared before the 1892 Royal Commission on Labour, also remarked that 'where there is a man and a family, who has no chance of making provision, there ought to be some arrangements made'.[63]

Finally, while minimising the dangers to themselves, working men spoke about the risks to women. Edwards, who noted the relatively recent entry of women into the dipping house, wanted the government to replace them with boys under eighteen. When asked if he made this request on behalf of women or men who didn't want women to work he replied:

---

[59] Transcript of arbitration, 9.
[60] Ibid. 15.
[61] Ibid. 13.
[62] Ibid. 107.
[63] Ibid. 146.

Of course I make my statement on behalf of the men, but my reasons for holding this opinion are simply these, that my experience in connection with this dangerous portion of our work had always been that young people and women have been more susceptible to the ill effects of lead poisoning than adult male labourers are.[64]

Not surprisingly, his request for the government to remove women from the dipping house elicited this question about possible trade rivalry. It also suggests that he was employing the dangerous label to support his argument for the exclusion of women from the most skilled and highest-paying job in the Potteries. This linguistic usage appears to be a tactic in the ongoing attempt of working men to, as Wally Seccombe has argued, realise the male breadwinner ideal through the displacement of their female trade rivals.[65]

Other potters had voiced similar requests during the years when the pottery trade was under investigation. For example, in their first communication with the Home Office in January of 1893, the Operative Potters' Association had written 'that women should not be allowed to work in processes in departments of labour that are proved to be unsuitable and imperious to them'.[66] Moreover, Edwards and other union officials spoke of the need to remove women from the dipping house and other lead-based processes because of their greater liability to lead poisoning during a government inquiry in 1908. Joseph Lovatt, general secretary of the National Amalgamated Society of Male and Female Pottery Workers, said that his organisation objected to women's employment in those processes because 'The women are not so likely to be able to withstand the inroads that lead makes upon them, and, from a humane standpoint, with regard to women who become mothers, their children suffer as a result of the mothers having worked in lead.'[67] When that group's president, Mr C., was asked if women should be excluded from those processes if the maintenance of a home was entirely dependent upon their wages, he replied:

I should hardly go so far as to say that, but, still, I think there are a great many girls, single girls, who could find occupation, such as you usually see advertised

---

[64] Ibid. 142–3.
[65] Wally Seccombe, 'Patriarchy stabilised: the construction of the male breadwinner wage norm in nineteenth-century Britain', *Social History* ii (1986), 53–76. Colin Creighton has critically evaluated the arguments of Seccombe and others on this subject in 'The rise of the male breadwinner family: a reappraisal', *Society for the Comparative Study of Society and History* xxxviii (1996), 310–37.
[66] Operative Potters' Association to Asquith, 19 Jan. 1893, HO 45/9851/B12393E.
[67] *Report of the departmental committee appointed to inquire into the dangers attendant on the use of lead: the danger or injury to health arising from dust and other causes in the manufacture of earthenware and china, and in the processes incidental therein, including the making of lithographic transfers*, PP 1910, [c.5385] xxix. 559.

in the daily paper.... When you tell them that they would be better in service, they tell you that they are not going to be anybody's floor-mop.[68]

He suggested replacing them with 'the boys that we often see standing at the corners of the streets with nothing to do, growing up idle men'.[69] This line of argument obviously piqued the committee's interest since it questioned Edwards about the employment of single women. He stated his opinion that single women's labour in the dipping house should be abolished because 'there were hundreds of them' with no one dependent upon their wages.[70]

These remarks are quite significant because the position of men in the pottery trade and their ability to realise the breadwinner ideal had been increasingly undermined since the 1860s. Marguerite W. Dupree has illustrated that increased employer reliance on female labour dated to the period when the extension of the factory acts to the pottery trade, along with the Education Act of 1870, resulted in the decreased employment of children. The proportion of women in the labour force rose from 31 per cent in 1861 to 38 per cent in 1881 to 45.6 per cent in 1901 and, as the number of women workers increased, they were disproportionately drawn from the families of potters.[71] They occupied assistant positions, formerly held by children, and moved into the clay departments that had been previously been male preserves. Dupree has concluded that these developments 'undermined the "breadwinner" family headed by male pottery-workers, as their relative wage-earning position gradually slipped in relation to other groups and the number and relative wages of women in the industry increased over the last third of the century'.[72] Richard Whipp has also made this point arguing that public references to the male breadwinner ideal were 'often the external image presented by certain men. Yet the private reality was quite different. Women were central, not marginal figures, in the territory which spanned between the household and factory'.[73] During their working careers, many women became the main or, sometimes, sole breadwinner due to fluctuations in the market or industrial disease among male family members. Both historians demonstrate the elusive nature of the male breadwinner ideal and the valuable monetary contribution women made to the household income.[74]

---

[68] Ibid. 584–5.
[69] Ibid. 585.
[70] Ibid. 587.
[71] Dupree, *Family structure*, 259.
[72] Ibid. 267.
[73] Richard Whipp, 'Work and social consciousness: the British potter in the early twentieth century', *Past and Present* cxix (1988), 142.
[74] For the importance of women's work to the trade and family economy see Marguerite W. Dupree, 'The community perspective in family history: the Potteries during the nineteenth century', in A. L. Beier, David Cannadine and James M. Rosenheim (eds), *The first modern society: essays in English history in honour of Lawrence Stone*, Cambridge 1989, 549–73, and Richard Whipp, 'Kinship, labour and enterprise: the Staffordshire pottery industry,

The disparity between this ideal and pottery men's actual position, however, did not prevent them from emphasising their role as the breadwinner and using that position as the basis for their demand for compensation during suspension. A group of workers met Chief Inspector of Factories Whitelegge on 22 March 1901 to put forward their idea of establishing an insurance fund, made up of contributions from employers and workers, for suspended workers. They were seeking Home Office support in the hope that the subject could be discussed during the arbitration. Whitelegge approved of this scheme, commenting that Home Office support 'would be well received by workers'.[75] The subject was discussed at arbitration and resulted in the establishment of a voluntary compensation scheme. Lord James ruled in 1903 that manufacturers could use any glaze recipe, providing they agreed to the medical examination of men and to provide compensation to lead-poisoned workers.[76] According to Peter Bartrip, eighty-six manufacturers joined the Pottery Insurance Company in its first year so that about half of the workers exposed to lead were covered by this scheme. Bartrip has argued that this outcome, 'the first ever compensation scheme for occupational ill-health', probably 'led to the 1906 extension of the Workmen's Compensation Act to cases of occupational diseases'.[77] A 1907 government order extended the provisions of that important legislation to workers suffering from lead poisoning.[78]

The proceedings and results of the arbitration reflect the way in which contemporary views about gender and work shaped the construction of measures to protect male and female pottery workers. Milner's exchange with Lord James succinctly encapsulates several salient aspects of contemporary views about masculinity and work. His comment, 'we take our risks on ourselves', reflects the fact that working men accepted personal risk and danger as a natural part of their working lives. And when their physical strength, a key indicator of their masculine prowess, was threatened by lead they practised self-help by taking voluntary and temporary leave from work

1890–1920', in Pat Hudson and W. R. Lee (eds), *Women's work and the family in historical perspective*, Manchester 1991, 184–203.

[75] Transcript of interview, Whitelegge and Messrs Hassell, Bennett, Parkes and Edward, 22 Mar. 1901, HO 45/10119/B12393P, 2.

[76] The contentious issue of manufacturers modifying their glazes to make them less toxic had been discussed during the arbitration. This ruling was essentially a compromise for those manufacturers who did not want to follow that course of action. It should be noted that the secretary of state had discretionary powers; he could insist upon the use of a leadless glaze if the compensation rules were violated or cases of lead poisoning occurred in a pottery: Bartrip, ' "Petticoat pestering" ', 21.

[77] Ibid. 23.

[78] *Order from Home Secretary Herbert Gladstone*, 22 May 1907, PP 1907, [c.3539] xviii. 767. The government's actions regarding the extension of workmen's compensation to diseases contracted at work were reported at the annual meeting of the TUC and a resolution for the inclusion of pottery work in workmen's compensation was unanimously passed: *Annual report of the TUC* (1907), 69–71, 166.

and collecting sickness benefits from their club. By the 1880s, according to Dupree, there was a wide array of registered friendly societies and sick clubs as well as unregistered sick clubs formed by male potters at their workplace. She has estimated that perhaps as many as 50 per cent of adult males in the Potteries belonged to one of these groups.[79] Clarie Holdsworth has argued that membership of such clubs, as well as paternalism among some employers, demonstrates the important tradition of self-help in the Potteries.[80]

The prospect of the state mandating a period of involuntary unemployment was an unwelcome one. Most obviously, as several working men remarked, it would undermine their position as the family providers. But there was more to their concerns, for, as John Tosh has suggested in his writing on masculinity, 'It makes a difference to recognise that unemployment not only impoverished workers but gravely compromised their masculine self-respect'.[81] Dr Arlidge had indeed argued against legislation for men years earlier, claiming that if employers and workers co-operated to improve the conditions of labour then 'the artificial bulwarks of Acts of Parliament will be uncalled for, and artisans will escape that sapping of self-reliance and independence'.[82] This and other comments made by doctors and government officials during the arbitration suggest a set of shared beliefs about men's work, including the belief that working men's status and sense of self-respect could be maintained by means of the provision of monetary compensation.

The perspective on state intervention in the lives of female pottery workers was dramatically different. Because of their role as potential mothers, they were subjected in 1898 to medical surveillance and suspension without consent or such monetary provisions. Thus, they were suspended for several years before compensation was available to them. Moreover, the continual discussion of possibly eliminating them from the lead-based processes of the trade, such as the dipping house, meant that they, unlike their male counterparts, faced the prospect of losing their jobs at any moment.

Why did the government chose not to take the step of banning women from the dangerous processes of the pottery trade? First, I would suggest that government action may have been influenced by the way in which lead was used in the trade. Workers were not exposed to raw lead, as in the white lead factories, but to lead in the pottery glazes or colours. Experimentation had also shown that manufacturers could reduce the threat of lead poisoning by producing glazes with less lead or even leadless glazes so theoretically there was greater flexibility in the processes. Second, and more significant, there

---

[79] Dupree, *Family structure*, 287–93.
[80] Holdsworth, 'Dr John Thomas Arlidge', 470.
[81] John Tosh, 'What should historians do with masculinity? Reflections on nineteenth-century Britain', *History Workshop* xxxviii (1994), 190.
[82] He made this statement in a lecture to potters in 1887: Holdsworth, 'Dr John Thomas Arlidge', 470.

was the diverse and extensive use of women's labour in this trade. Unlike the white lead trade, where women were concentrated in the white beds, thousands of women were diversely employed in this one. Moreover, they comprised nearly half of its workforce in 1901 and employers who relied extensively upon their labour strenuously opposed any suggestion of removing them. Unlike the white lead employers, they did not see women workers as easily replaceable; they wanted to keep them, with certain protective measures, despite the dangers of the trade. In order to safeguard their business interests, they repeatedly voiced their opposition to radical state action. Thus, I will argue that this case may be seen as another example of what Philippa Levine has called the 'uneven application of gender'. In her work comparing protective labour legislation and prostitution she has written, 'By closely defining the areas in which changes were to be made, the architects of these reforms ensured sufficient success to mitigate public anxieties without unduly alienating powerful business interests.'[83] As a result, the government created a compromise between the protection of women and their unborn children and the monetary interests of pottery manufacturers.

This was not the end of the matter, however, as concern over the impact of pottery work on reproduction led to further consideration of banning women. The subject was seriously discussed during the 1904 Committee on Physical Deterioration as well as at another departmental committee convened to investigate the conditions of labour in the Potteries in 1908. As will be seen in chapter 6, medical men were the chief proponents of this course of action. During the course of testimony in 1908, they buttressed their support for the exclusion of women from the lead processes by referring to the 'fact' that they were more susceptible to lead poisoning than men and, especially, that such work had 'an effect on them prejudicial to their childbearing functions'.[84] Despite such testimony, the committee concluded that these dangers could be greatly diminished or eliminated by the very strict observance of existing special or additional rules. In 1913 women were further prohibited from certain processes in the preparation of lead glaze, certain heavy work and work involving strain, without a certificate of permission to work.[85]

Medical theories about women, the issues of married women's work and reproduction, and women's place in the labour market were ever present and influential in the process of creating protective measures for the pottery trade. The regulation of this trade further significantly illustrates the emergence of a new way of talking about women's work, one that focused upon its harmful impact upon women and their unborn children. The discourse of reproductive danger represented a break from the discourse that guided the

---

[83] Philippa Levine, 'Consistent contradictions: prostitution and protective labour legislation in nineteenth-century England', *Social History* xix (1994), 33.
[84] *Report of the departmental commitee appointed to inquire into the dangers attendant on the use of lead*, 205.
[85] These rules, drawn up on 25 Aug. 1911, came into effect on 2 Jan. 1913.

making of protection in the 1840s and 1870s and had major implications for women workers. This new discourse and its implications will be discussed and analysed in the following chapters that explore more fully the role of the press, medical men and feminists in the making of dangerous trades regulations.

5

Narratives of Bodily Danger: the New Journalistic Press

> The cry of labour has seized the world's ear. The Press, the Legislature, and the world at large is listening to the voice of labour. . . . When this journal first resolved to secure a hearing for all working-class questions, there was scarcely a column of a leading London newspaper which was then open. Now, following our lead, every great daily paper has its labour section. . . . Nor is it only the press which is watchful. It is the readers of the Press.[1]

This self-promoting editorial in the *Star* in 1891 made a critical point: labour issues were becoming a standard feature in the daily London and provincial papers. As the making of dangerous trades regulations has shown, the press produced stories specifically about female labour in the nail and chain, white lead and pottery trades. Those mentioned in previous chapters represent only a fraction of the stories produced in the decades before the First World War when papers, such as the *Daily Chronicle* and *Star*, took the initiative and investigated a variety of trades employing women. The subsequent publication of information highlighting the special dangers of such trades for them, resulted in the creation of a series of industrial scandals.

This trend in reporting was part of the new journalism that developed in England between the 1880s and 1914. This period marked the rise of the mass daily newspaper with stories, often illustrated, written in a simple and lively style for its readers. These newspapers devoted great attention to the subjects of crime, sex, gossip and sport while de-emphasising the staples of older newspapers, parliamentary and political news.[2] The extensive and sensational coverage of the brutal Jack the Ripper murders in 1888 typifies the new focus in content. Further attempts to cater to the tastes of the mass audience included the creation of gossip columns, women's pages, children's features and comic pages.[3] Borrowing techniques from America, new journalists and editors, like William T. Stead and Thomas P. O'Connor, also produced interviews, *exposés* and political editorials in order to influence public opinion and promote what Stead called 'government by journalism'. Stead was responsible for what has been called the most successful piece of scandal journalism

---

[1] *Star*, 5 Sept. 1891.
[2] See Joel L. Weiner, 'How new was the new journalism?', in Weiner, *Papers for the millions*, 47–71.
[3] Ibid. 55.

in the nineteenth century, 'The maiden tribute of modern Babylon', which depicted young girls for sale to old men.[4] Written in 1885, it was critical to the passage of the Criminal Law Amendment Act that year, which raised the age of consent for sex from thirteen to sixteen.

Historians have analysed the various facets of the new journalism producing studies of leading journalists, newspaper owners, individual papers, their political orientations and interests as well as reader demographics.[5] Others have emphasised the increased and more sophisticated reporting on industrial issues.[6] They have, however, largely overlooked the proliferation of stories about women's work.[7] This chapter will examine this neglected but important development and answer two important questions: why did the press view women's work as a subject that would sell papers? What impact did the stories have upon women workers? I will contend that the press chose this subject because it was at the centre of contemporary cultural and political debates. Through their activities they were participating in the contest over women's work, and by extension, public space for them. Their participation in this contest also allowed them to pursue their agenda of 'government by journalism'; that is, promoting social reform and influencing the course of government action. In the end, I argue, they created narratives of bodily danger in the workplace that represented women as victims in desperate need of protection. These narratives, moreover, attracted both public and governmental attention and were critical in the enactment of dangerous trades regulations. Consequently, the press's efforts affected women workers through the establishment for them of legal boundaries, more radical than ever before, in the workplace and public space.

---

[4] Walkowitz, *City of dreadful delight*, ch. iii.
[5] The most valuable work is Weiner, *Papers for the millions*, but see also Brown, *Victorian news and newspapers*; Koss, *Rise and fall of the political press*; and Lee, *The origins of the popular press*.
[6] See John Goodbody, 'The *Star*: its role in the rise of the new journalism', in Weiner, *Papers for the millions*, 143–63. Brown has singled out the *Daily Chronicle* as an example of the new and improved industrial reporting of the 1890s in *Victorian news and newspapers*, 268–70.
[7] The few references to stories about women's work have not been found in literature specifically dealing with the developments within the press. For example, McFeeley has briefly mentioned stories about women's pottery work in *Lady inspectors*, ch. ix. Women's work in the match trade has received somewhat greater attention in Barbara Harrison, 'The politics of occupational ill-health in late nineteenth-century Britain: the case of the match industry', *Sociology of Health and Illness* xvii (1995), 20–41; Lowell J. Satre, 'After the match girls' strike: Bryant and May in the 1890s', *Victorian Studies* xxvi (1982), 7–31; and Walkowitz, *City of dreadful delight*, 76–80. Dina Copelman has examined the implications of the image of the sweated woman worker in the press in 'The gendered metropolis: fin-de-siècle London', *Radical History Review* lx (1994), 38–56.

'Government by journalism'

The most relevant journalistic trend for examination in this chapter is the emergence of the idea that the newspaper could and should function as an agent of social reform. William T. Stead and Thomas P. O'Connor were the leading editors to articulate this journalistic philosophy and try to make their respective papers, the *Pall Mall Gazette* and the *Star*, conform to it. Their influence, it should be noted, extended far beyond those two papers. In 1886 Stead published his controversial and often cited article 'Government by journalism' in the *Contemporary Review* in which he made great claims for the power of the press and editors.[8] The press had usurped parliament's power in the democratic age, he boldly claimed, and had become 'to the Commons what the Commons once was to the Lords. The press had become the Chamber of Initiative. . . . This new power of initiation it has secured by natural right'.[9] It was the locus of power in the democratic age; it would speak directly to the people. The editor, he argued, was the uncrowned king of democracy who wielded tremendous power. Each day he could administer either a 'stimulant or a narcotic' to the minds of his readers. If he chose the former, he could 'generate the steam, known as public opinion' which he claimed was the greatest force in modern politics.[10] Stead also avidly believed that the press could and should function as an agent of social reform. The newspaper, he asserted, 'has become what the House of Commons used to be, and still is in theory, for it is the great court in which all grievances are heard, and all abuses brought to the light of open criticism'.[11] It would serve as the 'great inspector' that uncovered abuses in places like prisons or workhouses and exposed them to public and governmental scrutiny. Finally, in order to fulfil its objective of 'government by journalism', he advocated the purposeful and cautious use of sensationalism. It was, Stead wrote, 'justified up to the point that it is necessary to arrest the eye of the public and to compel them to admit the necessity of action'.[12]

Thomas P. O'Connor's editorship at the *Star* (1888–91) was guided by similar ideas and owed a great debt to Stead and the *Pall Mall Gazette*. His article, 'The new journalism', published in the *New Review* in 1889, reflected his belief that a new style of reporting was needed for a contemporary audience. Living in 'an age of hurry and multitudinous newspapers', he noted, papers were often read in transit as people travelled to their destinations in railway carriages. In order, he wrote, to get your ideas into the 'whirling brains of your readers there must be no mistake about your meaning . . . you must

---

[8] For an excellent analysis of this work see Ray Boston, 'W. T. Stead and democracy by journalism', in Weiner, *Papers for the millions*, 91–106.
[9] W. T. Stead, 'Government by journalism', *Contemporary Review* xlix (1886), 656.
[10] Ibid. 661–2.
[11] Ibid. 673.
[12] Ibid. 653.

strike your reader right between the eyes'.[13] Comparing a newspaper to a street piano, he wrote that your 'notes should come out clear, crisp, and sharp'.[14] Disliking the 'old journalistic' attitude of independence and detachment from political issues, O'Connor advocated newspapers taking a political stand. 'A journal', he wrote, 'should be founded to advance definite and distinct principles, and should cleave to those principles.'[15] Asserting that a journal was a 'weapon in the conflict of ideas', he claimed that 'It should be used in accordance with the dictates of fair and honourable warfare; but it should be used as a weapon to wound the enemy and to defend the friend.'[16] The friend of whom O'Connor spoke was the mass of ordinary people. The *Star*'s first edition front page 'Confession of a faith' clearly stated its self-conscious designation as a people's paper. It would judge political policies according to their impact upon 'The charwomen who lives in St. Giles, the seamstress that is sweated in Whitechapel, the labourer that stands begging for work outside the dockyard gate in St. George's in-the-East'.[17] At its inception, then, the paper was concerned with appealing to ordinary people and covering issues relevant to their lives. This was evident as the paper reported on and supported one of the most significant labour events of the period: the match girls' strike.

'White slavery in London': the match girls' strike

In June 1888 Annie Besant's stirring article 'White slavery in London' was published in the *Link*.[18] It catalogued and publicised the industrial complaints of the exploited match girls working at the Bryant and May match company. Shortly after its publication and retribution, the dismissal of several workers to whom she had spoken, the female workers went out on strike and became the *cause célèbre* of London with extensive coverage of the strike appearing in numerous papers including the *Pall Mall Gazette* and *Star*.[19] As public opinion backed the match girls in their struggle, Bryant and May were forced to make concessions. Following their remarkable victory, the match girls formed a union. This strike is considered a pivotal event in the history of the new unionism and it also served, I will argue, as a prototype for later journalistic campaigns focusing on women's work issues.

Annie Besant shrewdly handled the match girls' story in the press. In her

---

13 T. P. O'Connor, 'The new journalism', *New Review* (1889), 434.
14 Ibid.
15 Ibid. 433.
16 Ibid. 434.
17 *Star*, 17 Jan. 1888.
18 Besant published articles in the *Link* on 23, 30 June, 7, 26 and 28 July, and 4 Aug. 1888.
19 The *PMG* published stories on 8, 14, 16 and 18 July 1888 while the *Star* wrote about the strike on 6, 7, 17 and 18 July 1888.

initial *Link* article, she vilified Bryant and May for its treatment of the 'helpless' girls. She related the bitter story of their forced contribution to a fund to erect a statue of William Gladstone. Many of the girls attended its unveiling with stones and bricks rather than sentiments of joyful commemoration. 'A gruesome story is told', she related to her readers, 'that some cut their arms and let their blood trickle on the marble paid for, in very truth by their blood.'[20] On a regular basis, their long hours, low pay and, especially, the fining system made it possible for the shareholders to make a handsome profit. 'Born in slums', she wrote,

> driven to work while still children, undersized because underfed, oppressed because helpless, flung aside as soon as worked out, who cares if they die or go on the streets, provided only that Bryant and May shareholders get their 23 percent . . .? Oh if we had but a people's Dante, to make a special circle in the Inferno for those who live on this misery, and suck wealth out of the starvation of helpless girls.[21]

A deputation of match girls to the Home Office on 14 July provided another opportunity to illustrate their pitiful plight. While they were not reenacting the Storming the Bastille, the spectacle of the group marching around the Embankment to the House of Commons attracted considerable attention. Besant's paper accentuated the physical differences between these 'ragged, gaunt East End' girls and the well-to-do West End folk. These workers were, another paper noted, 'pale, thin, undersized, ragged, their appearance eloquent of hard labour unfit for childish frames'.[22] That same day, the *Link* printed an open letter to the company's shareholders emphasising the incidence of phosphorous poisoning among the workers. It reported that the female hands developed this disease, commonly known as 'phossy jaw', because they ate their meals in the workrooms and phosphorous dust mixed with their bread. After describing how the poison ate away their jaw bone she posed a question to the shareholders: 'Do you not feel a twinge of pain in your own mouth as you think of these [girls] being poisoned that your table may be the more daintily spread?'[23]

Besant served as the 'people's Dante' as she presented this melodrama complete with its heroines, villains and moral. The match girls' oppressive story was told in the *Link* while the effects of dangerous work were also made visible through the spectacle of the ragged girls marching along the Embankment to the House of Commons. There was also a moral to the story for those interested in the plight of labour. As Walkowitz has argued, it demonstrated to contemporaries 'that with the aid of "sympathetic" press and public

[20] *Link*, 23 June 1888.
[21] Ibid.
[22] PMG, 14 July 1888.
[23] *Link*, 14 July 1888. The letter was also sent to the DC, the PMG, *The Times*, the *Daily News*, the *Daily Telegraph*, the *Star*, the *Echo* and the *Evening News*.

opinion, "the poorest and most helpless portion of the industrial community" could triumph over "the wealthiest and most powerful firms in the metropolis" '.[24] It further demonstrated effective tactics that could and would be deployed by newspapers interested in the conditions of women's labour. For Besant used techniques considered the hallmark of the burgeoning new journalism: investigative reporting leading to sensational stories that hit the readers 'right between the eyes', interviews and, especially, the use of the editorial as a political weapon.

At the conclusion of the strike and the formation of a union, the *Link* published its perspective on the idea of 'government by journalism'. The paper would strive to fulfil its larger mission of being the 'Word of the People' and do for other oppressed people what it has done for the match girls – voice their complaints and focus public attention upon their plight. 'Now, as of old', the editors wrote, 'it is true that the "dark places of the earth are full of the habitations of cruelty," and into those dark places *The Link* must be carried, let the outcry of those who prey in the darkness be as loud as it may.'[25] This was, the editors had previously remarked in an article on the press and the match girls, a difficult task. 'A political grievance', she wrote, 'felt by those who have the votes can be ventilated; but a grievance which presses on any weak and defenceless class, and which is inflicted by the rich and powerful, is rapidly boycotted by the press, with one or two honorable exceptions.'[26] This situation would change dramatically in the course of the 1890s as newspapers did extensively investigate and sympathetically treat labour issues.

The new journalism and dangerous trades

The theme of the physical dangers of women's work, secondary in the match girls' strike to exploitative working conditions, came to the forefront in extensive newspaper coverage of women's work in 1892 and 1898. During those years several newspapers, most notably the *Star* and *Daily Chronicle*, produced series of articles on hazardous work by special correspondents. The *Star* began 1892 by revisiting the match industry because it had uncovered cases of 'phossy jaw' among former Bryant and May workers. The *Daily Chronicle* ended the year with its highly shocking investigations of the white lead and pottery trades. In 1898 stories of death and illness in both the match and the pottery trades appeared once again in large numbers. As the previous chapters on dangerous trades regulations have shown, these pieces attracted the attention of the reading and political public and ultimately made their way into newly-created Home Office files for government investigation.

---

[24] Walkowitz, *City of dreadful delight*, 77.
[25] *Link*, 4 Aug. 1888.
[26] Ibid. 28 July 1888.

They were seminal in the emergence of this new avenue of protective labour legislation and the subsequent designation of all three trades as dangerous trades. Now I will examine more carefully how they reported on the subjects.

In the aftermath of the infamous strike, Lowell Satre has argued, Bryant and May emerged as a model company with respect to its technology and its treatment of its workers.[27] The company's paternalism was reflected in numerous ways including support for the Clifden Institute, founded after the strike, which provided evening classes, a club room, lodging for a small number of workers and meals at a low price. The company, moreover, placed no obstacles in the way of membership of the union or of friendly societies. But, as Satre notes, the union born out of the strike posed very little challenge to the company. This positive perspective on Bryant and May would be shortlived, however, as the *Star* uncovered cases of 'phossy jaw' among some of its former employees.

On 18 January 1892 readers of the paper were told about the 'Star Man's' incidental discovery of 'phossy jaw' at Bryant and May in an article entitled 'The "phos": a terrible disease that scourges the matchmakers'. He told the story of a woman who began to experience tooth and jaw aches after five years of work. The company doctor sent her home and then extracted four teeth but, she told the reporter:

> That didn't do no good. Then the pain got worse. . . . It was just as if somebody had got something scraping the bones in my cheek. And then he said my husband and the children must not be in the same room with me, because the smell was so bad. The doctor went for his holidays, and while he was away lumps of my bone worked right out through my cheek – it was festering dreadful. I kept the bone to show him, but it smelt so awful I had to throw it away.[28]

When the doctor returned and reexamined her, he pronounced her cured and payment from the company, one pound for thirty-one weeks, stopped. Despite being 'cured' she was unable to find work again because she had the 'phos'. 'It frightens the other girls', she explained, 'and besides if you get it a second time, it generally kills you.'[29] As a result, she was disfigured and her family was having financial troubles. 'For six years', the writer editorialised, 'she has HELPED TO SWELL THE DIVIDEND of Bryant's shareholders, now she starves.'[30] And she was not alone: 'All these tales of misery and suffering are within gunshot of each other; and "the plague is in every house," either actual sufferers, or knowing of cases. The whole district where the matchmakers live is an abode of rottenness, all due to the "phos".'[31] Thus began the paper's campaign against the company.

---

[27] Satre, 'After the match girls' strike', 13–15.
[28] *Star*, 18 Jan. 1892.
[29] Ibid.
[30] Ibid.
[31] Ibid.

The *Star* reporter's visits to other match factories and interviews with current employees led him to blame Bryant and May for the spread of this plague. The workers revealed that the company was negligent in the provision of soap, water and towels to clean off the phosphorous dust. When asked about the provision of these items one worker replied 'What, soap? No Fear! Give you nothing. If we want soap we've got to buy it – they give us sand and half cold water. And we only got that a month or two ago.'[32] These conditions, one worker volunteered to the reporter, were not as bad as being 'choked' by the fumes emanating from the dipping room. Because of the factory's arrangement, the girls had to walk through the dipping room in order to obtain their frames. To explain what this entailed the journalist described the work processes for the readers:

> In the process of manufacture the splints of wood, twice the length of an ordinary match, are packed in rows in a square frame, so that both ends project. The phosphorous is then spread out on a flat surface and kept to a uniform thickness by an iron gauge. Into this mixture the two ends of the splints referred to above are dipped, and then these framed are hung in the drying room. The room in which the operations are performed is called the dipping room, and the fumes of the phosphorous can be seen ascending from the phosphorous as it lies upon the surface.[33]

Unlike the workers, the manager he interviewed did not think there was a problem or any need to answer questions about illness in his factory. He told the reporter:

> We do not see that we are called upon to sit here, and be interviewed by newspapers. We have NOTHING TO SAY TO THE STAR or any other paper. We have had enough experience of their unfairness in the past. They go and put in any lie they hear, and we are supposed to reply.[34]

The writer argued that simple reforms could minimise the effects of the disease: rearrangement of the factory so that the workers would not have to pass through the dipping room and the provision of supplies for personal cleanliness.

On 25 January the paper introduced medical testimony into its crusade against Bryant and May. It printed the graphic details of the progression of the disease as described by Dr Gant who told of toothaches, the loosening of teeth with puss exuding from their sockets, followed by swelling of the gums and face. If only one side of the jaw was infected, he noted, the person's face would look peculiarly lopsided. Dr Gant concluded that 'Death may, at length, take place from exhaustion, or rapidly from gangrene of the cheeks

---

[32] Ibid. 19 Jan. 1892.
[33] Ibid.
[34] Ibid.

and lips; recovery occasionally ensues, with some considerable loss of bone and deformity.'[35] In common with all other authorities, the writer asserted, Dr Gant argued that the 'disease arises exclusively from the fumes of common phosphorous'.[36] This opinion substantiated the paper's recommended preventive measures.

As the scandal was gaining momentum, the *Star* appealed to the company's shareholders.[37] A column published the day before their semi-annual meeting tried to instill in them a sense of shame and responsibility for the scourge of 'phossy jaw'. The column writer made a particular appeal to the female shareholders whom he addressed as 'mothers, wives, and daughters like the wretched match girls themselves'.[38] You can no longer, he wrote, 'close your eyes to the sad offspring of misery of which you are parents! How long for pity's sake, shall the rotting bones of the match-girls cry out to you in vain for mercy?'[39] Following this appeal to gender solidarity, the paper concluded its final plea with the last stanza of a poem entitled 'Lucifer in the East', which had appeared with its first article:

> Ay! go where the brimstone has shed its blight,
> Where its victim is cowering thro' day thro' night,
> With an agony such as no tongue may speak,
> Shunned as a leper, and anguished and weak,
> Penniless, workless, forlorn.
> Go, shareholder, you with the dividend fair,
> Go, see, and consider it well,
> How the daughters of women are perishing there,
> In you Lucifer's Brimstone Hell![40]

Despite this adverse publicity, the company did not initiate any reforms in 'Lucifer's Brimstone Hell'.

While disappointed by the company's denials and inactivity, the newspaper could fairly report in June that their articles had produced belated and limited results. For, in the wake of its revelations, the Home Office conducted its own investigation of Bryant and May, declared the match trade a dangerous trade and issued special rules on 2 June. The company was required to reorganise its buildings to isolate the dipping room and improve its ventilation systems. Most significantly, workers in any match company complaining of a toothache or swelling of the jaw were supposed to be examined by a

---

[35] Ibid. 25 Jan. 1892.
[36] Ibid.
[37] According to Satre, the company's chairman made no reference to the reported cases of illness. When questioned by a shareholder who made reference to the *Star*, he replied that it would be 'very undignified for us to notice any articles' and assured them that 'your factories are model factories': 'After the match girls' strike', 18.
[38] *Star*, 27 Jan. 1892.
[39] Ibid.
[40] Ibid.

doctor immediately. If they exhibited the symptoms of phosphorous poisoning, he had to report the case to a factory inspector.[41] In this way, the state could monitor the health of workers in this dangerous trade.

The *Star*'s response to these developments exhibited a mixture of shameless promotion and criticism of the government. Its 8 June front-page editorial emphasised the paper's considerable efforts in exposing this horrible industrial scandal. It had extracted testimony 'at no small expense of pains' from reticent ex-employees, it had shown the dangers to which young girls were exposed and it had 'appealed for justice for those poor toilers whose unsightly, rotting faces, might have moved the most stony-hearted to pity'.[42] They had failed to move the shareholders, however, but 'Late in the day – almost five months late – the HOME SECRETARY has realised the cogency, in law and in fact, of the *Star*'s contentions.'[43] While they could thus claim victory, they were critical of the state's tardiness in recognising that this was a dangerous trade. 'Englishmen must tingle with humiliation', a 10 June article told *Star* readers, when they knew that they lagged behind other countries, including Austria, Germany, Switzerland and Denmark, in the enactment of regulations to safeguard the match workers.[44]

In the end, the *Star* narrated two stories. In addition to telling the story of women who were victims of industrial illness it recorded the course of its literary campaign. The quote at the beginning of the chapter illustrates its self-perception as stimulating journalistic interest in labour issues. The paper's investigative efforts regarding phossy jaw were the subject of front-page self-congratulatory editorials. For example, on 25 January they noted:

> Correspondents' letters and the comments of some of our contemporaries – papers in some instances in strong antipathy to the political aims of *The Star* – show us that these unimpeachable revelations of hideous suffering are touching the chords of pity common to ALL CLASSES AND PARTIES. But we must do more than expose the wrong. We must continue to cry aloud until the wrong is righted.[45]

The *Star* furthered its claims to influence through the inclusion of extracts from medical journals referring to its articles. Under the heading, 'Listen to the "British Medical Journal" ', they quoted at length from a recent article in that prestigious journal. Its writer noted that

---

[41] For details of the Home Office investigation see Harrison, 'The politics of occupational ill-health', 20–41.
[42] *Star*, 8 June 1892.
[43] Ibid.
[44] Ibid. 10 June 1892.
[45] Ibid. 25 Jan. 1892.

Some recent articles in the *Star* newspaper on this subject [phosphorous necrosis of match workers] have, we must confess, taken us by surprise; for, after making due allowance for frequent incorrectness of articles on medical topics in non-professional journals where medical supervision of the matter is not exercised, there still remains evidence that the horrible disease . . . still haunts the occupation of matchmaking.[46]

That same day, the newspaper quoted from a reader named 'Humanity' who was grateful for 'the manner in which you have brought before the public the nature of this terrible disease'.[47] The correspondent claimed gratitude on behalf of the match girls and all people who called themselves Christians. In this manner, the paper emphasised its power of initiating a dialogue on the subject and eliciting support from diverse readers, cutting across class and political lines, to create a consensus that the government must do something about this scourge among the match workers.

Lead poisoning and infant slaughter

The *Star's* approach and interest in highlighting the grave results of women's work in a dangerous trade was not unique. This book has already introduced another major paper, the *Daily Chronicle*, which unveiled the findings of its special commissioner regarding women's work in the white lead trade. Consequently, lead poisoning joined phossy jaw as an industrial disease on the journalistic agenda. The mention of legalised infanticide in the white lead trade, as chapter 3 has shown, created quite a stir in public and political circles. And, by bureaucratic standards, the problem was resolved rather quickly when in 1898 women were prohibited from the most dangerous sectors of the trade. Using the same journalistic techniques and approach, the *Daily Chronicle* then pursued the lead-poisoning problem in the pottery trade. As it turned out, the story led to less dramatic results but even more extensive media coverage.

After the initial series of articles on lead poisoning in the Potteries in 1892, the newspaper's reporting was minimal until the spring of 1898. Then, in May, the paper's special representative revisited Staffordshire in the aftermath of revelations in the local paper, the *Staffordshire Sentinel*, of blindness among pottery girls. His first article entitled 'Blind by the lead: a visit to the pottery girls' appeared, interestingly juxtaposed to an article and picture of female frocks and fashions. With minimal commentary he related the blind workers' pitiful stories and claimed 'that if a house-to-house canvas of the Potteries were undertaken it would be found that not a quarter of the cases

---

[46] Ibid. 28 Jan. 1892.
[47] Ibid.

have been brought to light'.[48] Five days later he published an article with a scathing critique of the special rules enacted in 1894. Arguing that they were not stringent enough to combat the pernicious lead, the writer advocated the introduction of leadless glazes into the industry. He knew of at least one pottery factory that had employed this method of production to cut down on lead poisoning.[49] Consequently, he pressed for the government to appoint a committee of specialists to experiment on glazes. Employers could be encouraged to work toward this objective, he believed, by designating lead poisoning as an injury entitled to compensation outlined in the 1897 Workmen's Compensation Bill. 'Premiums on raw lead', he claimed, 'would be simply prohibitive.'[50] The articles' proposals were in line with those made before Home Secretary Ridley in the highly publicised deputation discussed in chapter 4.

Pressure for government action also appeared frequently on the paper's editorial page. In March 1898 it commented upon the discussion of the problem in the House of Commons. It noted that 'The House has taken the case of the blind girls to heart. . . . The only matter before the House was the ghastly scandal of the pottery trade, with its unspeakable trial of death and suffering and impoverished lives.'[51] In anticipation of the May deputation, the paper seized upon a report that the Home Secretary was going to refuse to see the blind and crippled workers travelling from the Potteries. The paper believed that this decision was based upon the fear that the workers were to 'be paraded and made a spectacle for sensation's sake'.[52] Yet it could authoritatively state that 'Every member of the deputation comes prepared to tell the Home Secretary the conditions of work under which hundreds of thousands of English lives had been crippled, ruined and destroyed every year, to answer questions, to speak from personal experience.'[53] The newspaper writer, moreover, had interviewed them and pronounced them to be 'intelligent persons, competent to describe and discuss those conditions'.[54] In other words they were credible witnesses.

The *Daily Chronicle* unanimously condemned the Home Secretary's refusal to meet the workers and his decision to appoint yet another investigative committee and enact further rules. In an article entitled 'Rose water for the plague', the author characterised Ridley as 'a weak, squeezable, Minister' who 'requires to be rescued from his own feebleness, and to be provided with some artificial substitute for a backbone and to be pushed on the side of humanity

---

[48] DC, 14 May 1898.
[49] Ibid. 19 May 1898.
[50] Ibid.
[51] Ibid. 5 Mar. 1898.
[52] Ibid. 19 May 1898.
[53] Ibid.
[54] Ibid.

harder than some employers push him on the side of profit'.⁵⁵ The side of humanity, which included the paper, required him to do his duty towards a trade in which

> the lives of men, women, and children are treated as dust in the balance, that young men and women who, if a little care were taken, would grow up strong and healthy, are poisoned, blinded, crippled, subjected to excruciating tortures, and then tossed on the side, that the scourges of phthisis rampant in the workshops, ALL THIS WASTE OF LIFE AND PROLONGED AGONY OF SUFFERING IS NEEDLESS, because the poison and dust can be rendered harmless by a moderate amount of caution and an insignificant expenditure of money.⁵⁶

The editor hailed the Home Secretary's proposal to enact new special rules as 'rose water for the plague'. Instead, he wrote, the appointment of a female factory inspector and the exploration of the possibility of using leadless glazes, would be far stronger measures than the rose water cure.

Other papers followed in this vein of criticism. The *St James's Gazette* felt that 'The principle of appointing women as inspectors has been thoroughly established elsewhere; and here, if anywhere, is an appropriate field for the best work at their command.'⁵⁷ They characterised Ridley's actions as a compromise 'which will hardly be accepted with the equanimity that so frequently smile over the formalities of red tape'.⁵⁸ In an article entitled 'White lead martyrs', the *Star* endorsed 'with all the indignity of our common humanity', the *Chronicle*'s indictment of the Ridley. 'We know', the paper editorialised, 'that the House of Commons is cold, callous, slow of ear, deaf of heart, but we cannot believe that it will ratify this monstrous inhumanity, the commercial cruelty, to which the Home Secretary is consenting.'⁵⁹ Likewise, a further editorial entitled 'Potter's rot and Ridley' charged him with indifference to people whose stories of suffering 'read like a canto out of the "Inferno"; more horrible than even Dante imagined'.⁶⁰

In other scathing editorials, the press particularly lambasted the Home Secretary for his refusal to meet the group of six sick pottery workers who had travelled to London expressly to meet with him. They also used pictures and sketches of the blinded workers to prove the Home Secretary's callousness. The *Daily Chronicle*, the *Penny Illustrated Paper* and the *Westminster Gazette* for example, included these visual aids to assault the readers' sensibilities and to portray Ridley as a real villain. 'If there were any doubts', the reporter for the latter paper wrote,

---

55 Ibid. 21 May 1898.
56 Ibid. 19 May 1898.
57 *St James's Gazette*, 23 May 1898.
58 Ibid.
59 *Star*, 21 May 1898.
60 Ibid. 30 July 1898.

as to the terrible effects of lead poisoning on men and women working in the Potteries, these would be quickly removed by a glance at the group from which the accompanying sketches were made yesterday. The men are both paralysed either in hands or legs, while the women are all three more or less blind. Sir Matthew Ridley did not think it necessary to see the sufferers personally[61]

The *Daily Chronicle*'s caption detailed the ailments of each poor lead victim noting that 'Mrs. Martin, the eldest woman in the group, was working on a dinner service for the Queen when she was seized with blindness.'[62] The *Penny Illustrated Paper* carried the caption 'Charles Dilke champions the lead poisoned workers of the Potteries'. Describing the ailments of the workers pictured, the writer noted 'One of the most shocking features of lead poisoning is the high percentage of children who are attacked by the lead. . . . It is quite certain that until the use of raw lead is prohibited there will be a repetition of these hideous cases.'[63] Ridley certainly made a tremendous blunder when he refused to meet the workers.

This was not the end of the story, however, as papers sought to maintain the tempo of this scandal. Graphic stories of illness and death in the Potteries became standard fare in the daily papers. Significantly, the campaign that had begun with the stories of blinded workers had evolved to promulgate the increasingly familiar theme of the impact of lead on women and their unborn children. The *Daily Chronicle*, for instance, ran a series entitled 'Lead in the home' in June 1898. In the first instalment, 'Whole families desolated: a story of mothers and children', the writer recounted his journey into the 'houses in these mean streets' where women and men spoke of family life in these towns 'with their devastating infant mortality and their terrible trade diseases'.[64] The women interviewed conformed to a pattern: once young and healthy they prematurely aged because of the breakdown of their health. For example, Mrs Jackson married at twenty-one and was then 'strong, handsome, and a very well formed girl, as her photograph shows'.[65] After six years of cleaning the ware and occasionally carrying glazed tiles on a plank over her back, she was poisoned by the lead. After the onset of lead poisoning she was never able to speak clearly again. He described another woman, Mrs Hulme, as a very handsome woman 'but her features are deeply lined, and her face has the concentrated and terribly sad expression that comes from constant pain'.[66] Her physical ailments included leg, head and eye problems. Meanwhile, Mrs Basford could not stand for any period of time and 'her legs are swollen, her hands are weak and trembling, and her speech is almost

---

61 *Westminster Gazette*, 20 May 1898.
62 *DC*, 6 June 1898.
63 *Penny Illustrated Paper*, 4 June 1898.
64 *DC*, 18 June 1898.
65 Ibid.
66 Ibid.

incomprehensible'.[67] As a result her bricklayer husband had to translate her story, which included her difficulties bearing healthy children. While employed, she had given birth to two children, one of whom died nine months after birth, and had suffered two miscarriages.

'Infanticide in the Potteries: weeping for her children because they are not' embellished upon the lack of thriving families with more instances of women and, especially their children, 'lost to the lead'. Mrs Bourne was 'healthy as a girl, and is still a handsome young woman of that robust but stately and gentle type of beauty that seems common in the Potteries'.[68] She developed lead poisoning after five years work in the dipping house. Although she stopped working and her first baby was healthy, her last two died within months of their birth. The writer then described Polly Bates whose mother died of lead poisoning at the age of forty. Polly . . .

> an elfish shred of a creature, nine years old, is a curious specimen of a lead worker's child – the feeblest, frailest little being, with a long, thin stalk of a neck, and a long thin head. She is quite blind, and there is something more than human about the extraordinary vivacity of the piping voice and fumbling fingers belonging to a body which hardly looks substantial enough to stand a puff of wind.[69]

There was no need, the writer remarked, to seek a 'specific connection between her blindness and the mother's occupation'.[70]

This ghastly theme of infanticide in the Potteries was further developed in a column, 'Lead poisoning from day to day', which began the following month. Each day the *Daily Chronicle* bombarded its readers with detailed cases of illness and death as well as poignant personal tales of tragedy. On 4 July 1898, for example, the column further discussed medical opinion on the effects of pottery work on pregnancy. It quoted from a recently published work by Dr Prendergast entitled *The potter and lead poisoning*. It reproduced his statistics from France showing that in '123 pregnancies [of lead workers], 73 children were born dead . . . of the fifty born alive twenty died in the first year, eight in the second, seven in the third, one later, and only fourteen reached the age of ten'.[71] The article concluded with the now well-known Dr Oliver's opinion that 'Lead workers miscarry in much larger proportion than other women; and the children born generally die of convulsion.'[72]

The *Daily Chronicle*'s 'human interest' approach not only reflected the new trend in reporting industrial illness but generated much publicity and income for the paper itself. Like the *Star*, the paper repeatedly emphasised the

[67] Ibid.
[68] Ibid. 27 June 1898.
[69] Ibid.
[70] Ibid.
[71] Ibid. 4 July 1898.
[72] Ibid.

publicity their reports had generated in self-congratulatory editorials and carefully noted their multiple audiences. For example, a June editorial ran 'We are glad to see that English women of all degrees are coming to the help of the pottery workers' in reference to resolutions passed by the WLF and WCG.[73] The editorial added

> Conservative and Liberal opinion in the House and in the Press is at one with Mr. Morley's view that the facts which we have helped bring to light – and of which alas we still have a heavy unpublished record – scandalise and horrify the national conscience. Certainly it will not be our fault if the rescue movement for the Potteries takes on a party complexion.[74]

Their rescue mission, as they termed it, was discussed in other papers. They noted that the *Bradford Observer* had referred to their articles and said:

> it is almost like another chapter from Defoe. Instance after instance is given of young women of fine natural physique disfigured and tortured through a shortened life by loss of sight, speech, strength, mental facilities – worse than that, giving birth to children who are old and decrepit at birth, or breaking down in the office of maternity.[75]

They were even more pleased that the most important provincial paper, the *Manchester Guardian*, endorsed their efforts to bring to light what was going on in the Potteries.[76] The attention their articles and editorials received certainly increased the paper's commercial value.

Unlike the *Star*, however, the *Daily Chronicle*'s articles stimulated a different kind of dialogue with the pottery manufacturers who vociferously challenged the press's representation of the special hazards of the trade through letters to the editor and a counter-narrative in the *Pottery Gazette*. For example, pottery manufacturers Burgess and Leigh critiqued the paper for unfairly using its powers and portraying all manufacturers as callous and indifferent to the worker's plight. Their defence rested upon the claim that workers were careless and bore responsibility for the incidence of lead poisoning.[77] These and other themes were developed far more substantially in their paper in the spring and summer of 1898. For instance, an article of 1 April put forward their view of the nature of their trade. While admitting that girls and women in the dipping house are often pale and unhealthy, they claimed 'that labour as a whole is unhealthy, always has been unhealthy and will continue to be unhealthy so long as one portion of the people have to do too much to enable the other part to do little or none'.[78] They contrasted this

---

[73] Ibid. 23 June 1898.
[74] Ibid.
[75] Ibid. 30 June 1898.
[76] Ibid.
[77] Burgess and Leigh, pottery manufacturers, to the editor of the *DC*, 27 May 1898.
[78] *PG*, 1 Apr. 1898, 449.

naturalised view of the dangers with the pronouncements of 'hysterical writers' who were inflaming feelings with their hyperbolic stories.[79] Their favourite example came from a writer who wrote of the 'holocaust of human beings offered up annually at the altar of the glaze of porcelain'.[80] In August the journal reprinted an article from a Manchester newspaper which claimed that 'The party journals are still playing to the democratic gallery of the theatre of public opinion. This time it is the pottery trade that is being attacked'.[81] The *Pottery Gazette* agreed and added their belief that the offensive newspapers pursued a policy of 'baiting' employers 'for the edification of trade unionists, Democrats, and Socialists'.[82]

As seen in chapter 4, pottery manufacturers, unlike their counterparts in the match and white lead trades, lobbied the government against some of the proposals put forward by the coalition of labour and the press. This counter-attack played a role in the press's limited success in achieving their goal of government by journalism in the Potteries. For while the *Daily Chronicle* publicised the horrific lead-poisoning problem, found common cause with labour and convinced other newspapers of the necessity for more stringent measures, it ultimately failed to achieve its objectives. The government enacted new rules in 1898 but they did not include either the appointment of a female factory inspector or the introduction of a harmless leadless glaze.

The preceding stories about women's work are just a fraction of those produced by the new journalistic press in the decades before the First World War. I would suggest that they bear some similarities to stories that Walkowitz has called 'narratives of sexual danger'. Perhaps not as overtly sexy as the purveyance of young girls chronicled in the 'The maiden tribute of modern Babylon', they none the less narrated an equally significant story of sexual danger. The *Daily Chronicle* and the *Star* significantly contributed to, and reinforced, the idea that women outside prescribed social boundaries faced danger. In this case, they conveyed the message that women faced bodily harm in the workplace. Match workers were vulnerable to disfiguration and death because of phosphorous poisoning. Moreover, as Dina Copelman has argued, the image of the sweated woman worker (such as the match workers) in the press conveyed an important message that

> at the heart of the Empire, in the world's most advanced capital, was not just an anonymous working-class aggregate, but exploited and stunted female workers, hindered from fulfilling their duties as present or future wives or mothers, because of their miserable work conditions.[83]

---

[79] A more sustained criticism of the press was found in the *PG*'s series of articles published on 1 June, 1 July, 1 Aug. and 1 Oct. 1898.
[80] *PG*, 1 Apr. 1898, 449.
[81] *Textile Mercury*, 9 July 1898, reprinted in *PG*, 1 Aug. 1898.
[82] Ibid.
[83] Copelman, 'The gendered metropolis', 45.

This link between work and the inability to fulfil their spousal and maternal duties was made more explicitly in the sensational stories of women and foetuses perishing in hazardous lead trades. As the lead gained 'access under all coverings, to every part of the body', it preyed upon women's reproductive organs, the most important parts of their bodies. It thwarted their reproductive abilities and endangered any unborn children. The physical dangers of this labour for women and their potential offspring emerged as the common theme of the various articles.

Moreover, these stories were like the 'Maiden tribute' in the sense that they created a moral panic.[84] Walkowitz has outlined this phenomenon as

> the definition of a 'threat'; the stereotyping of main characters in the mass media as particular species of monsters; a spiraling escalation of the perceived threat, the taking up of absolutist positions, including the mounting of the 'barricades': and finally, the emergence of an 'imaginary' solution, in terms of tougher laws, moral isolation, and symbolic court action'.[85]

The press's treatment of the hazards of women's work in three trades reflects this process. The stories were simple and melodramatic; Bryant and May were the 'monsters' of the match trade while Home Secretary Ridley was the villain of the pottery scandal. All women, in contrast, were portrayed as potential victims of industrial illness as their bodies were threatened by poisons. This distorted representation of the problem, moreover, was the key to stimulating public outcries for the protection of these visible and vulnerable women and their endangered offspring. The *Star* and *Daily Chronicle*, in particular, erected 'barricades' with their aggressive investigations that sought to instigate government action. In so doing, they became a leading force in the extensive extra-parliamentary discussion of possible state intervention in the lives of working women.[86] Their stories were mentioned in questions in parliament and, most important, they made their way into Home Office files. Home Office papers indicate a direct link between newspaper publicity and the initiation of government investigations. With the resulting new measures they could claim, to varying degrees, to have fulfilled their mission of representing the interests of a marginalised group, that is women who lacked political power and agency. In so doing, they promoted 'government by journalism'.

The press purposely selected women's work as a topic of investigation

---

[84] The subject of moral panic has also been discussed in Nan H. Dreher, 'The virtuous and the verminous: turn-of-the-century moral panics in London's public parks', *Albion* xxix (1997), 246–67, and Erich Goode and Nachman Ben-Yahuda, *Moral panics: the social construction of deviance*, New York 1994.
[85] Walkowitz, *City of dreadful delight*, 121.
[86] This development substantiates James Vernon's point about the centrality of print culture to late nineteenth-century politics: *Politics and the people: a study in English political culture c. 1815–1867*, Cambridge 1993, ch. iii.

because it was not only a labour issue but a source of significant social anxiety. It seized upon an issue which could transcended geographical, class and political boundaries and would sell papers. One irritated pottery worker suggested this explanation in an article in the *Staffordshire Sentinel* in 1909. 'The Potteries', she wrote, 'are a butt for outside agitators. If they wish to say or write something sensational, they turn to the Potteries for a subject.'[87] A few weeks previously, she noted, a Socialist MP went there to agitate and recorded after his visit that he 'had seen girls unsexed- aye, almost dehumanised-by the conditions of their industry'.[88] She was trying to repel the attacks on women's work, especially married women's work, which were reaching a feverish pitch in the years before the First World War. Chapter 1 has shown that anxiety about women's work was not new; from the first years of industrialisation anxious observers had condemned women working in factories or in jobs previously considered men's work. Such labour violated the prevalent and powerful Victorian ideal of separate spheres for men and women. This ideal retained its force over the course of the nineteenth century but was expressed in different ways at different times. As I have shown in chapter 1, opponents of female labour in the 1830s and 1840s claimed that it led to pre- and extra-marital sex, sexual exploitation and illegitimacy. During the 1870s they emphasised that it caused women to neglect their duties to their children often leading to infant mortality. By the decades preceding the First World War, the dangers which were described and dramatised were the hazards of work to their bodies or, more specifically, to their reproductive organs. The press's narratives produced during the 1890s were a critical component in this re-conceptualisation of a long-standing social problem.

Very significantly, men were conspicuously absent from the resulting moral panics despite the fact that they, too, worked in these officially designated dangerous trades. For instance, although women outnumbered men in the match trade, men worked in its most dangerous processes: the mixing of paste and the dipping and drying of matches. Most females worked at cutting down and boxing.[89] Male contact with lead was significant, especially in the pottery trade, where they dipped the pottery in the leaded glazes. Newspaper articles made some mention of male potters, as in the case of the famous 1898 deputation, but for the most part the journalistic and public gaze fell upon the women. Moreover, although when men replaced women in the white beds of the white lead trade in 1898 and government papers documented a significant rise in illness and death among them, the press did not choose to

[87] SS, 21 July 1909.
[88] Ibid.
[89] In 1895, for instance, Bryant and May employed about 2,000 workers, between 1,200 and 1,500 of them female. According to statistics from the annual report of the chief inspector of factories for 1898, this was the overall trend for all match companies. This point has been made in Harrison, 'The politics of occupational ill-health', 22.

publicise this disturbing development. In no instance whatsoever did the impact of phosphorous or lead upon the male worker's ability to produce healthy children appear in stories about infant slaughter. In the press, as in medical writings and government investigations, danger in the workplace was specifically linked to one sex.

Gender figured in the production of these stories as well as in their emphasis on the endangered woman. During the 1890s the staff of the *Star* and *Daily Chronicle* was male.[90] Between 1888 and 1891 Henry W. Massingham, George Bernard Shaw and several Fabian Socialists including Sidney Webb wrote for the *Star*. A tangible link between the two papers, Massingham joined the *Daily Chronicle* in 1891 and served as its editor from 1895 to 1899. He was a radical journalist with an avid interest in labour politics and policies and was, for a brief period, a member of the Fabian Society. In his 1892 work, *The London daily papers*, Massingham quoted a politician who told him that 'The most influential paper in this country is the *Daily Chronicle*'.[91] Massingham boasted that the paper had the most extensive labour coverage of any of the daily papers and was 'probably nearer the inner mind of the Left wing of the Radical Party than anything which figures in the *Daily News*, the *Pall Mall Gazette*, or the *Star*'.[92] Vaughan Nash served as the paper's labour expert during Massingham's tenure as editor.[93] Co- author, with Herbert Llewellyn Smith, of a book on the 1889 London Dock Strike, Nash wrote the paper's articles on the deadly trades.[94]

In the end, the press created stories which could be read on multiple levels. Most obviously the *Star* and *Daily Chronicle* told the story of silent women workers who suffered because of their work in hazardous trades. They simultaneously narrated their own story, emphasising their leading role in uncovering and publicising this information as well as their efficacy in engaging a diverse groups of readers. The public and governmental responses to their tales also provide the historian with insight into prevalent social concerns about women's work and potential motherhood. The implications of these narratives of dangerous work for women were immense. Women were banned from working in the most dangerous and highest-paying parts of the white lead trade and were subjected to medical examination and possible suspension from the pottery trade. In addition to these official results there were unanticipated ones. As noted in chapter 4, the press's portrayal of the Potteries as the source of baby slaughter prompted some manufacturers to take the initiative and replace female with male workers. In other instances

---

[90] For this information I have relied primarily upon John Goodbody, 'The *Star*', and Alfred F. Havighurst, *Radical journalist: H. W. Massingham, 1860–1924*, London 1974.
[91] Henry W. Massingham, *The London daily press*, New York 1892, 121.
[92] Ibid.
[93] Havighurst, *Radical journalist*, 20.
[94] Evelyn March-Phillipps identified him as their author in 'Factory legislation for women', 743.

women colluded with their employers to keep their health status quiet. Both approaches, McFeely has contended, were engineered to avoid employers being called 'baby killers'.[95] The narratives also affected women workers in trades other than those discussed in this chapter, for they ensured that phrases such as 'dangerous trades' or 'unhealthy trades' entered into the common public vocabulary of pre-war protection. Frequently and variably used, they were appropriated by different groups, invested with different meaning and intention, in a variety of situations. Hereafter, translating the words 'unbecoming', 'indecent', 'unfit', to 'unhealthy', 'dangerous', 'hazardous', was a powerful strategy for those who wanted to limit women's working opportunities. This rhetorical strategy will be seen more fully in the next two chapters which focus on the role of doctors and feminists in the public discourse on the dangers of women's work.

---

[95] McFeely, *Lady inspectors*, 65.

6

Medical Men, Sexual Science and Dangerous Trades Regulations

In 1895 Dr John Arlidge wrote a short piece in the *Journal of the Sanitary Institute* on the study of industrial diseases. It was not hard to review the literature for, as he correctly noted, 'These disorders have received slight attention . . . and in consequence the literature on the subject is meagre.'[1] But, he quickly added, that had all changed recently as industrial diseases had received tremendous attention because of governmental inquiries into the causes of sickness and death in certain trades. 'Moreover', he wrote, 'like other novel movements, the subject has been ardently taken up by the "Press".'[2] While acknowledging the press's role in drawing attention to the subject, he nevertheless criticised its partisan perspective, inaccuracies and exaggeration. He claimed that its sensational stories, meant to stir up public opinion, were the fruit of 'uneducated observation'. In contrast, he and other medical men who had been studying industrial illness for decades were the experts who could provide reliable information about the disease and sound advice on preventive measures.[3]

Arlidge was writing at an important moment in the history of industrial illness. There was, as he suggested, an unprecedented interest in the subject and this was reflected in the myriad special investigations and the creation of dangerous trades regulations. All of this activity represented a dramatic shift in government policy, a shift from a largely *laissez-faire* approach to industrial illness to one of cautious prevention. At this juncture, medical men came forward with their specialised knowledge and figured prominently in the white lead and pottery proceedings. As the creation of regulations for those trades has shown, Dr Oliver was particularly significant for his service on several dangerous trades committees and for undertaking special investigations on behalf of the Home Office. Arlidge, as well as Dr Thomas Legge, the first medical inspector of factories, were among the multitude of doctors who provided testimony or their written works on the subject of dangerous trades.

---

[1] Dr John Arlidge, 'The position of the study of industrial diseases: its past neglect and its scope', *JSI* xv (1895), 517.
[2] Ibid.
[3] In 1892 Arlidge published *The hygiene, diseases, and mortality of occupations*. According to Clare Holdsworth, this work was the first significant British work published on occupational diseases since 1831. For more on Arlidge's contribution to the study of occupational illness see Holdsworth, 'Dr John Thomas Arlidge', 458–75.

There was a new awareness of the dangers of lead work and this was particularly significant for women workers. For, according to these authorities, lead poisoning was a 'woman's problem'. As chapters 3 and 4 have shown, medical men advanced the theory that women were more susceptible to lead poisoning than men. They also predominantly linked the idea of dangerous work to women and their reproductive abilities. Men did not figure in the discussion of the hazardous white lead trade at all and only in a minor way in the pottery trade. Even then there was no mention of the harmful impact of lead on their virility or ability to produce healthy children.

Barbara Harrison's work has pointed to several problematic aspects of this gendered diagnosis of the lead-poisoning problem. Medical men, she has written, had difficulty in 'separating the effects of work from the total environment in which women lived, and the diseases of work from the diseases of poverty'.[4] As a result 'it was common to find [in their accounts of industrial women workers] the concept of "susceptibility" used to refer to an environmental rather than a biological predisposition to the effects of industrial conditions'.[5] She has further challenged their concept of suspectibility because of their methodology; crude rates of poisoning were accepted as indicative of a greater disposition to lead poisoning without consideration of controls such as age, processes of work or the proportion of women to men employed in a trade.[6] Regarding the critical issue of lead and miscarriages she has written that 'Lead was commonly used as a abortifacient and there seems to have been a common perception of its qualities. But this raises another difficulty for establishing the validity or the possible risks of lead poisoning. High rates of miscarriage were in any case commonplace among working-class women.'[7] Thus, it is difficult to ascertain the precise relationship between lead work and the incidence of infant mortality. Her comments raise very valid and important questions about medical evaluations of lead poisoning.

This chapter will examine more extensively the influential medical writings on lead poisoning produced between 1830 and 1914, with particular emphasis on those written between the 1890s and the First World War. My analysis reveals that the emphasis on sexual difference as the key determinant in the development of the disease was a new and questionable theory emerging in the 1890s. It is most significant that the writings themselves contain evidence that bad conditions in the workplace contributed to the

---

[4] Barbara Harrison, 'Women's health or social control? The role of the medical profession in relation to factory legislation in late nineteenth-century Britain', *Sociology of Health and Illness* xiii (1991), 476
[5] Ibid.
[6] Idem, ' "Some of them gets poisoned": occupational lead exposure in women, 1880–1914', *Social History of Medicine* ii (1989), 181–2.
[7] Idem, *Not only the 'dangerous trades'*, 90. This subject has also been discussed in Patricia Knight, 'Women and abortion in Victorian and Edwardian England', *History Workshop* iv (1977), 57–69.

incidence of the disease and that men were equally liable to illness and death. Despite this information, Oliver and others still maintained that lead poisoning was a 'woman's problem'. I will argue that this conceptualisation of the lead-poisoning problem was shaped by their understanding of sexual difference and concern about the future of the race. They were sexual scientists who believed that biology was destiny, that nature outfitted women to be mothers not workers. Moreover, their writings suggest that they were motivated by an overriding concern to protect unborn children, 'the future of the race', and for this reason they may be viewed as the chief advocates of what we today call 'foetal protection'.

Medical perspectives on the pottery and white lead trades, 1830s–late 1880s

According to Anthony Wohl, a fatalistic attitude towards industrial illness prevailed during the nineteenth century:

> Industrial diseases were simply an accepted part of working life, as inevitable and as unpleasant as the long hours of work or uncertainty of employment. Miner's asthma, or, as it was more graphically called, 'black spit', potter's asthma, or 'potter's rot', . . . – these and many more were all part and parcel of the Victorian vocabulary.[8]

There were, however, a few progressive medical men who did not share this complacent attitude and wrote about the dangers of certain trades. In 1831 Charles Turner Thackrah published a study of occupational diseases in which, A. Meiklejohn has argued, he made 'a unique contribution to medicine' with his idea that preventive measures should be taken to safeguard the health of workers in various trades.[9] Thackrah noted that workers in all departments of white lead works were thin and sallow and complained of headaches and loss of appetite. But the men working in the white beds and packing departments were most adversely affected by their labour. They

> soon complain of head-ache, drowsiness, sickness, vomiting, griping, obstinate constipation, and to these succeed colic or inflammation of the bowels, disorders of the urinary organs, and, finally, the most marked of the diseases from lead, palsy. We observed the muscles of the fore-arm more frequently and sooner to suffer than other parts. The eyes are also affected with chronic inflammation, or reduced nervous power.[10]

[8] Wohl, *Endangered lives*, 264.
[9] A. Meiklejohn, *The life, work and times of Charles Turner Thackrah, surgeon and apothecary of Leeds (1795–1833)*, London 1957.
[10] Charles Turner Thackrah, *The effects of the principal arts, trades, and professions, and of civic states and habits of living on health and longevity: with suggestions for the removal of many of the agents which produce disease, and shorten the duration of life*, London 1832 edn; repr. London 1957, 103.

He believed that more than half the illness in this trade could be prevented if workers changed clothes, washed and brushed their hands and skin after work, and bathed regularly. The men should not be allowed to eat their meals in the workrooms which, he further noted, ought to be spacious and well-ventilated. Thackrah also discussed the incidence of constipation, colic and paralysis among men who dipped pottery into glazes. After inquring into various possible remedies, he asserted that the 'total disuse of lead in glaze is highly desirable' to prevent injury to those workmen.[11]

Dr John Simon, medical officer to the privy council, directed the systematic investigation of a variety of industrial diseases linked to certain trades, including the pottery and white lead trades, in the mid-nineteenth century. The previous chapter has mentioned the writings of Dr Edward Greenhow and Dr Arlidge on the pottery trade. Their commentary, it is interesting to note, suggested that respiratory illnesses were the main threat to pottery workers' health, not lead poisoning. Dr Whitely conducted an inquiry into the white lead trade and the results were incorporated in Simon's 1863 report for the privy council. As Simon commented:

> for all the operatives scrupulous personal cleanliness greatly diminishes the danger; and that even the most endangered persons – those, namely, whose employment lies in an atmosphere containing considerable quantities of lead-dust, and who get poisoned rather through their mouths than through their skin, might, by the special precaution of using respirators, avert much, perhaps nearly all, of whatever suffering now results from their occupation.[12]

He concluded, however, that the dangers in the white lead trade were 'evidently too great to be entirely within control of personal cleanliness, and perhaps too great to be in any way altogether remediable, while the present process of manufacture continues. It is alleged that the dangers have been diminished by the improvement of ventilation and worker cleanliness'.[13] Both Whitely and Simon agreed that the trade deserved further investigation.

At this point, the literature on lead poisoning was also sparse. Dr James Alderson, the premier man in the field, discussed the effects of lead on workers in his Lumleian Lectures delivered to the Royal College of Physicians in 1852 and reprinted in *The Lancet*. 'It is a fact now so clearly understood', he said 'as to become familiar, that when lead, either in fumes or powder, is diffused in the atmosphere, it may be absorbed into the system. . . . The effects of the imbibed poison are no means constant; different individuals are affected in dissimilar ways.'[14] He ascribed the variable symptoms and devel-

---

[11] Ibid. 121.
[12] Dr John Simon, *Public health reports*, London 1887, 107.
[13] Ibid.
[14] Dr James Alderson, 'On the effects of lead upon the system', *Lancet* mdccclii (1852), 75.

opment of lead poisoning to 'the peculiarity of individual constitution or of the especial mode of absorption of poison'.[15]

There were, then, few references in this period to lead poisoning in the pottery and white lead trade. And in each instance sexual difference was not mentioned as a factor contributing to the development of pulmonary diseases or lead poisoning nor were women singled out as particularly vulnerable to the dangers of these trades. And although medical reports contained numerous comments about the link between women's work and infant mortality, these two trades were not singled out as especially dangerous in this respect. Simon and the others produced these investigations to draw governmental attention to the issue of industrial illness. They hoped that by illustrating the relationship between conditions in certain trades and the incidence of illness they could instigate state actions. The government responded to their findings with modest provisions for the potteries in 1864 and for white lead works in 1883. The response was modest because of the prevalent fatalistic attitude towards industrial illness – an attitude shared by government, employers and workers.

Lead poisoning is a woman's problem, 1891–1914

Governmental and medical activity during the 1890s reflected a major departure from the earlier perspectives on the hazards of the lead trades. The government vastly expanded the amount of information it collected on the conditions of workplaces and the incidence of illness. It devoted tremendous time and effort to investigations, designated the trades as dangerous and created special measures to protect workers. Now it was not just a few progressive medical men who were interested in combatting the ill-effects of this work but rather, as noted in the introduction, a whole corps of medical men including MOHs, CSs, the chief and medical inspectors of factories. Yet, at the same time as the scope of governmental involvement was dramatically increasing, the focus of its investigations was becoming more narrow. For instance, the pottery trade was primarily viewed from the perspective of lead poisoning rather than pulmonary illness resulting from dust inhalation. Most significant for both trades, sex, or rather sexual difference, was propelled to the forefront of the discussion of the dangers of lead work. Doctors argued that women were more liable to the disease than men and paid unprecedented attention to women's work and its relationship to infant mortality. Oliver and his avowedly novel theories about lead poisoning were most influential in these critical developments.

The subsequent lectures were also published in that volume at pp. 165–7, 212–14, 391–3, 416–19.
[15] Ibid.

In the spring of 1891 Oliver delivered the highly prestigious Goulstonian lectures before the Royal College of Physicians of London. His subject was lead poisoning and because, as he acknowledged, his opinion was 'so totally at variance with that given by several authors' he explained it in some detail.[16] It was the opinion when Sir James Alderson delivered the Lumleian Lectures in this college in 1852, he said, 'that women were less frequently affected than men; but such is not my opinion. Women suffer far more frequently and severely than men'.[17] He attributed women's greater liability to lead to 'sexual idiosyncrasy'. It has been shown, he noted, that given equal exposure, as in epidemics of lead poisoning from lead in the water supply, 'the fair sex is weaker and that their nerve centres are more easily undermined by lead'.[18] His experience at the Royal Infirmary in Newcastle confirmed that perspective as the number of women suffering from lead poisoning outnumbered men. He provided statistics showing that between 1883 and 1888 ninety-one women and forty-four men were treated for that ailment.[19] Among women, he further noted, there was a variation in the development of the disease. The 'class' of women readily affected, he asserted, were the ill-fed, the badly housed and thinly-clad girl or women who labour to 'support idle or drunken husbands, or paramours, or who have lost their husbands'.[20] Tempted by the high wages, they became the victims of lead poisoning. In his final lecture, Oliver addressed the crucial linkage between lead work and reproductive disorders. 'Lead as a poison', he said, 'strikes early at the functions of blood-making and reproduction, producing sterility, liability to abortion, amenorrhoea or menorrhagia. Woman, from her constitutional idiosyncrasy, therefore, is more liable to be impressed by lead.'[21] Again he would say that lead 'disturbs the utero-ovarian function. . . . Lead workers miscarry in a much larger percentage than other women'.[22]

Oliver's lectures reached a wider medical audience when they were published in the *British Medical Journal* that same year. His opinions were also publicised and popularised in sensational newspaper stories in the *Daily Chronicle* in December 1892. They generated tremendous public and governmental interest in a subject previously confined to a small number of medical men. Oliver's contention that women should be removed from the dangerous parts of the white lead trade was seriously considered by the 1893 White Lead Committee appointed to investigate the trade. Following testimony from numerous doctors supporting his new theory, linking the sex of the worker to

---

[16] Dr Thomas Oliver, 'Goulstonian lectures on lead poisoning in its acute and chronic manifestations', *BMJ*, 14 Mar. 1891, 572.
[17] Idem, 'Goulstonian lectures', ibid. 8 Mar. 1891, 507.
[18] Ibid.
[19] Idem, 'Goulstonian lectures', 14 Mar. 1891, 572.
[20] Ibid. 573.
[21] Ibid.
[22] Idem, 'Goulstonian lectures', ibid. 21 Mar. 1891, 630.

Table 2
Lead poisoning in the white lead trade, 1898

Month	Males	Females	Total
January	14	31	45
February	22	14	36
March	13	24	37
April	14	19	33
May	18	28	46
June	21	9	30
July	28	9	37
August	31	5	36
September	67	1	68
October	38	2	40
November	34	1	35
December	46	1	47

Source: Oliver, 'Lead and its compounds', 296.

risk in lead work, the government felt compelled to protect these endangered women by eliminating them from their dangerous work. This was, by bureaucratic standards, quickly and easily achieved in 1898.

This was not, however, the case when Oliver and his co-investigator prescribed a similar ban on female pottery workers that same year. The government did not endorse that course of action and, in fact, Chief Inspector of Factories Whitelegge argued that the proposal only added to the 'vexed question of women's work'.[23] Most significantly, he doubted Oliver's theory of women's greater susceptibility to lead poisoning because of the publication of statistics on increased illness and death among the men who had replaced women in the white beds of the white lead trade. As previously noted, in Newcastle-upon-Tyne eighty-two males compared with twelve females were ill in the six months following the removal of women from the dangerous processes of the trade and this trend continued. Many medical men still believed that lead poisoning was a 'woman's problem' even when they were confronted with this and other statistical evidence illustrating that the men who replaced the women in the white beds were becoming ill and dying in greater numbers.

This point was clearly made by Oliver in his chapter 'Lead and its compounds', contributed to the massive volume he edited in 1902 entitled *Dangerous trades: the historical, social, and legal aspects of industrial occupations as affecting health, by a number of experts*. He incorporated national statistics for the white lead trade, as well those for Newcastle-upon-Tyne, and both illustrate the rise in illness among male white lead workers. Table 2 replicates his table of recorded cases of lead poisoning in the white lead trade, drawn

---

[23] Whitelegge minute, 2 Mar. 1899.

Table 3
Lead poisoning: in-patients, Royal Infirmary, Newcastle-upon-Tyne, 1892–1900

Year	Total	Recoveries		Deaths		Remaining on books
		M	F	M	F	
1892	44	15	27	2	1	2
1893	32	5	25	–	–	2
1894	31	7	20	–	–	4
1895	35	11	18	1	–	5
1896	38	12	22	–	2	2
1897	21	7	12	1	1	–
1898	36	22	12	–	–	–
1899	20	19	1	–	–	–
1900	14	14	–	–	–	–

Source: Oliver, 'Lead and its compounds', 298.

from the annual report for 1898 of the chief inspector of factories. Oliver commented that 'This table shows the transference of the incidence of plumbism from female to male operatives.'[24]

The same trend was demonstrated in Oliver's table (table 3) of statistics regarding the treatment of white lead workers at the Royal Infirmary, Newcastle-upon-Tyne. Here Oliver not only noted the dramatic decline in illness among females but also the fact that 1900 represented 'the first time in the history of the Newcastle Infirmary within our memory a whole year passed without even one female being received'.[25] While admitting that the foregoing statistics might raise doubts about the greater susceptibility of women to lead poisoning, he added 'Admitting for the moment that the susceptibility is equal in the two sexes, and the fact, too, that in both the illnesses may be severe, still I unhesitatingly assert that in main the symptoms are neither so severe in men, nor does the malady run so rapidly to a fatal termination as it does in women.'[26] Once again he attributed this outcome to women's physiological makeup when he wrote 'To a certain extent the reason is to be found in the fact that lead exercises an injurious influence upon the reproductive functions of women.'[27]

Oliver also buttressed his position by referring to a 1901 article published by Dr Legge entitled 'Industrial lead poisoning'. Legge discussed government statistics, gathered in 1899, of poisoning among men and women in different lead trades. Certifying surgeons reported the overwhelming preponderance of

---

[24] Oliver, 'Lead and its compounds', 296.
[25] Ibid. 299.
[26] Ibid. 298. Oliver repeated this point in *Diseases of occupation from the legislative, social, and medical point of view*, London 1908, 150.
[27] Ibid. 301.

Table 4
Lead poisoning in the pottery and white lead trades, by sex, 1899

	Severity of symptoms				
	Severe	Moderate	Slight	Not stated	Total
Pottery trade					
Male	79	22	232	7	340
Female	3	–	17	2	22
White lead trade					
Male	35	16	62	6	119
Female	22	18	67	4	111

Source: Legge, 'Industrial lead poisoning', 100.

lead poisoning among male workers: 954 cases among them compared to 176 cases among female workers during that year. Of this total, 340 men and 22 women were reported ill in the white lead trade while 119 men and 111 women were reported ill in the pottery trade. The breakdown of those figures according to the severity of symptoms by sex, in table 4, reveals that a higher number of severe cases were reported for male workers.

After presenting this numerical data, Legge discussed the issue of the greater susceptibility of women to this disease. He remarked that the 1899 lead-poisoning figures 'do not contribute much to the solution of the question as to whether or not women are more susceptible to the effects of lead than men . . . and the figures support the conclusion that the proportion of severe attacks is greater in males than in females'.[28] He then tried to answer the question of susceptibility by examining the statistics of illness and death in the white lead trade in Newcastle-upon-Tyne after women were banned from its lead-based processes in 1898. These figures, he conceded, 'do not lead to the conclusion that females are more susceptible to lead poisoning than males'.[29] However, he, like Oliver, immediately added 'the influence of lead on the child-bearing function is of immense importance'.[30] He then proceeded to use data from the women factory inspector's report of 1897 and an inquiry conducted by Dr Arlidge. The latter had collected information about the number of children born to 239 women before, during or after their employment in the pottery trade. His statistics illustrated a higher incidence of miscarriage when women worked in the trade. Legge's review of the available literature led him to note that the number of miscarriages to pregnancies was calculated to be between one in five to one in ten.[31]

[28] Legge, 'Industrial lead poisoning', 103.
[29] Ibid.
[30] Ibid. 103–4.
[31] Ibid. 105.

It is interesting to note that Legge's concession that men and women could be equally liable to the development of lead poisoning was only a temporary one. He would, for instance, testify before a government committee in 1908 that women were indeed more prone to the development of the disease.[32] As late as 1934 he wrote that the 'returns of occupiers do not lend themselves readily to a solution of the problem, as no census is given of the number of persons exposed to risk in the different processes.... Still, I consider women to be more susceptible to the effects of lead poisoning than men'.[33] In conclusion, then, if the first medical argument was questionable, opponents of women's work in this dangerous trade could buttress their position by referring to the impact of lead on their unborn children.

The Potteries revisited

The tenacity of Oliver's questionable theory, as well as the linkage of women's work to infant mortality, are reflected in the continual discussion of the dangers of women's work in the pottery trade. Medical men, the chief purveyors of these two perspectives, played a key role in keeping the discussion on the subject alive from 1898 until the outbreak of the First World War. For instance, they spoke about it at professional meetings and wrote about it in publications in 1898 when the government was considering its ban on women. Their opinions are found in government investigations including the 1904 departmental committee on physical deterioration, women factory inspector's reports for 1906, and the 1908 departmental committee examining the dangers of lead and dust in the pottery trade. A few examples from those proceedings will illustrate doctors' continuing belief that drastic action was necessary to protect the unborn children of women pottery workers.

While the government challenged Oliver's authority and decided not to remove pottery women, many of his medical colleagues supported his and Thorpe's recommendation. For example, in his address before the Sanitary Institute in 1899, Percy Frankland praised their 'courage' in dealing with the difficult subject of prohibiting women from the leaded parts of the pottery trade. This recommendation, he argued further, 'will serve to show future generations that in the 19th century there were, at any rate, some who recognised, however dimly, the direction in which the most important of all hygienic reforms would some day be carried out'.[34] There was no doubt in his mind, he told his audience, 'that women are more susceptible to lead poisoning than men and that plumbism is attended with particularly disas-

[32] *Report of the departmental committee appointed to inquire into the dangers attendant on the use of lead*, 421.
[33] Dr Thomas Legge, *Industrial maladies*, London 1934, 64.
[34] Dr Percy Frankland, 'Address to section iii', *JSI* xx (1900), 390.

trous consequences in the matter of producing healthy offspring'.[35] It was, moreover, their duty to protect offspring; to secure conditions so that women, and especially mothers, could 'live and carry on the most important industry of all – the production of men'.[36] In his 1898 publication, *The potter and lead poisoning*, Dr W. Dowling Prendergast asserted that 'Women are more amenable to the action of lead than men.'[37] He wrote of lead's destructive action on 'red blood corpuscles' leading to anaemia, as well as its impact on menstruation and gestation. He supported the exclusion of women, especially young girls from the dipping house, because of such effects. 'Her uterine and nervous functions', he wrote, 'are quickly vitiated. If married, maternity is denied them. The history of many working women in contact with the lead is a continued series of abortions.'[38] He argued that men could easily replace such at-risk women in such dangerous work. Both before and after the publication of Oliver's report, the *British Medical Journal* reiterated what was becoming the familiar perspective. 'Nearly all female workers in lead', an 1898 article asserted, 'are either menorrhagic or amenorrhoeic, and abortions are so common as to materially lower the birth rate in some of the affected districts.'[39] The greater susceptibility of women to lead poisoning mandated their removal from jobs that brought them into contact with the dangerous substance. Following the publication of Oliver's report, the journal wrote that 'Until the problem of the fritting of lead compounds has been satisfactorily solved, it is clear that safety lies in excluding from the dangerous processes those who by sex and age are predisposed to lead poisoning.'[40]

Medical opinions figured prominently in the 1906 annual report made by the principal lady inspector, Adelaide Anderson. Including in her report an array of statistics illustrating high infant mortality in the trade, she drew the inescapable conclusion that they were 'undoubtedly largely due to the extensive employment of married women in the china and earthenware trades'.[41] She buttressed this general statement with a series of quotations from MOHs working in the pottery districts in Staffordshire. Dr Dawes, MOH for Longton, said:

> There is, undoubtedly, a regular slaughter of innocents every year in Longton due to this [taking their children to nurses and exposing them to inclement weather] and premature births, and I am afraid that until we can obtain legislation to prevent mothers from going out to work we shall not be able to reduce these two cause of infantile mortality.[42]

---

[35] Ibid. 389.
[36] Ibid. 390.
[37] Dr W. Dowling Prendergast, *The potter and lead poisoning*, London 1898, 30–1.
[38] Ibid. 46.
[39] 'Lead poisoning in the Potteries', BMJ, 23 July 1898, 245.
[40] 'Lead poisoning in the Potteries', ibid. 6 May 1899, 117.
[41] *Report of the chief inspector of factories and workshops for 1906*, PP 1907, [c.3586] x. 234.
[42] Ibid.

Dr Clare, MOH for Hanley, concurred, noting that there was an increased tendency to employ women. In fact, he found many instances where the husband didn't work and the wife was the breadwinner. The MOH for Fenton, Dr Hughes, also linked the existence of female breadwinners to the infant mortality problem. 'Any attempt', he said, 'to combine the office of child-bearer and breadwinner in one person must of necessity result in feeble, bottle-fed babies and premature births.'[43] In an interesting concluding remark, Anderson suggested that working men had anticipated these medical arguments sixty years previously.

The case against women's pottery work was made even more strenuously in testimony before the 1908 departmental committee appointed to inquire, yet again, into the dangers of the trade. And, once again, the exclusion of women from the lead processes of the trade was one of its primary considerations. Appearing as a witness rather than as chief medical advisor, Oliver merely restated his long-held opinion that women and young girls should be prohibited by virtue of the fact that they were more susceptible to lead than males and because 'they are the mothers of future people'.[44] Not surprisingly, most of the medical men questioned made similar remarks. For example, Dr Legge, Dr Prendergast and Dr Alcock, who had made their objections to women's pottery work known in previous forums, repeated them before the committee.[45] Prendergast supported the exclusion of women from the lead processes because 'it is most deadly and most deleterious to the life of the women and their children'.[46] Alcock testified that he already 'excludes them when they have a history of repeated miscarriages and pregnancy – the fifth and sixth month'.[47]

However, Dr George Reid, the committee's medical expert and the MOH for Staffordshire, deviated from the consensus. While he concurred that lead work adversely affected pregnant women he did not entirely endorse the current proposal. For, he argued, this might lead the women to seek employment in other branches of the trade where they might suffer even more severely, that is, in the dusty processes. He presented statistics of higher mortality rates due to dust inhalation than lead poisoning. Moreover, he targeted the period after birth as the most perilous one for infants. Early return to work after confinement, he argued, diminished the child's chances of survival because they were deprived of breast feeding. In the end, he recommended the extension of the mandatory four-week leave after childbirth to three months.

---

[43] Ibid. 253.
[44] *Report of the departmental committee appointed to inquire into the dangers attendant on the use of lead*, 765.
[45] Other lesser known medical men followed suit. See the testimony of Dr John Russell at p. 432, Dr Arthur Hill at p. 472 and Dr Herbert Folker at p. 487.
[46] Ibid. 437.
[47] Ibid. 455.

Reid was essentially restating a position he had adopted in the 1890s after he had undertaken an investigation of the high infant mortality rate in the pottery district. The subject first came to his attention, he noted, in a paper presented to the Sanitary Institute in 1894, when he collated the reports of various MOHs for Staffordshire County Council.[48] He was struck by the disparate rates of infant mortality for the northern and southern towns and connected them to the different work opportunities available to married women. They found plentiful employment in the Potteries of the north while the iron and coal trades in the south offered limited work opportunities. He attributed the higher infant mortality rate to women's employment. For, he noted, they left their children in the care of others and, most important, they were fed artificial foods rather than being breast-fed. He believed that the state had to intervene to protect these young children and reiterated these points in all his subsequent works. For instance, Reid wrote in 1902 that the current four-week prohibition after childbirth 'is valuable from the point of view of the mother's health, it can hardly benefit for the child, for, if factory work is to be engaged in after a month's interval only, it is not likely the mother will commence suckling her child'.[49] Four years later, at the National Conference on Infantile Mortality, he asserted that 'endless proof is available of the injury to infant life arising from the unrestricted practice of mothers engaging in factory labour'.[50] If women were restricted from working for three months, he said, at least they could nurse their children during that time so that 'the most precarious time period of the infant's life would be tided over, and the chances of subsequent survival would be considerably enhanced'.[51]

All the medical witnesses, then, linked women's work in the trade to infant mortality. While they agreed on this critical point they disagreed about the causal relationship. Reid represents what might be called the 'old school' as his remarks hark back to those made by medical men in the 1870s. They had focused upon the women's neglect of their maternal duties; they abandoned their children for the mills and deprived them of maternal care, good food and nursing. The current discussion of this persistent and unresolved problem continued to centre around the issue of bad mothering and infant mortality but with a different emphasis. Beginning in the 1890s, medical men expressed the belief that certain work was physically dangerous for women and their unborn children. The act of child-bearing rather than child-rearing became the central focus; the period of pregnancy rather than the first year

---

[48] His paper, 'Infant mortality and female labour in relation to factory legislation', was published in the *JSI* xv (1895), 497–503.
[49] Dr George Reid, 'Infant mortality and factory labour', in Oliver, *Dangerous trades*, 84–9 at p. 88.
[50] Dr George Reid, 'Infant mortality and the employment of married women in factory labour before and after confinement', in *Report of the proceedings of the National Conference on Infantile Mortality*, London 1906, 223–36 at p. 223.
[51] Ibid. 229.

after birth was seen as the dangerous period. In this way, they were narrowing down the interpretive perspective: motherhood in its more limited biological aspect replaced the larger social definition. This represented a key shift in the conceptualisation of the long-standing woman worker problem which was being further developed in the writings of a new group of 'progressive' medical men.

Into the womb: ante-natal concern

The writings of Dr George Newman, who served as the chief medical officer at the Board of Education in the years before the First World War, exemplify this reorientation among some medical men. Newman discussed, at great length, the harmful effects of lead work in a chapter entitled 'Ante-natal influences on infant mortality' in his 1906 volume *Infant mortality: a social problem*. After referring to Oliver's illustration of the baneful effects of lead on foetuses, he argued that mothers should be the focus of the state's preventive measures:

> Infant mortality in the early weeks of life is evidently due in large measure to the physical conditions of the mother, leading to prematurity and debility of the infant; ... it becomes clear that the problem of infant mortality is not one of sanitation alone, or housing, or indeed of poverty, as such, but is mainly a question of motherhood.[52]

From his perspective, environmental factors such as sanitation or poverty had only a secondary effect upon the unborn child. Since its early development was dependent upon the mother she most directly affected its physical condition. And because motherhood was of 'vital importance to the nation', Newman pressed for efforts to achieve a higher standard of 'physical motherhood'.[53]

Dr J. W. Ballantyne, physician to the Royal Maternity Hospital in Edinburgh, even more emphatically asserted the need to focus on ante-natal conditions. In his presentation at the 1906 National Conference on Infantile Mortality he asserted that life began in the months in the womb not at birth. In order, then, to obtain the greatest benefits from preventive medicine they had to 'begin the prevention with the beginning of life; in other words, we must try to protect the as yet unborn infant'.[54] He reiterated this point in numerous works including his 1914 *Expectant motherhood: its supervision and*

---

[52] Dr George Newman, *Infant mortality: a social problem*, London 1906, 257.
[53] Ibid.
[54] Dr J. W. Ballantyne, 'Ante-natal causes of infant mortality, including parental alcoholism', in *National Conference on Infantile Mortality* (1906), 126. Ballantyne devoted tremendous attention to the development of the foetus and its relationship to its mother in a section entitled 'The ante-natal economy'.

hygiene. 'A new discovery calls for a new commandment', he claimed, in reference to the discovery of the 'health value' of the nine months that preceded birth. The purpose of this work was to investigate the various factors which might affect the infant in the womb, including the care and protection of the mother.

Ballantyne spoke specifically about dangerous work for women and its impact on their unborn children. 'I do not think', he wrote, 'the risks of causing abortions which certain trades carry with them are sufficiently recognised and guarded against by women expecting to become mothers. Workers in lead, such as type-founders and pottery-glazers, are especially liable to abortion, premature labours and dead-births.'[55] Such work was particularly threatening to a nation which was experiencing a falling birth rate. He advocated the more rigid closing of such trades to pregnant women. 'If women must work for their living', he wrote in 1914, 'while they are bearing children for the replenishment of the stock of babies, the Government must see to it that they are enabled to rest, at any rate, from the labours which are dangerous to their unborn infants as well as themselves.'[56] Unborn children were even more valuable now, he wrote, 'due to the diminished output in babies (speaking again in somewhat commercial language)'.[57] Most particularly, he said 'there has come to be a concentration of interest upon the expectant mother because of her expectant infant'.[58] He himself acknowledged that his motives might appear 'mercenary' but acting upon them would produce the desired increase in the national birth rate.

Various MOHs joined these two men in advancing the campaign against infant mortality along this new line of argument. Dr Cooper Patten, MOH for Norwich, was one of many public health officials calling attention to the need to attack the falling birth rate by concentrating upon the 'antecedent conditions unfavourable to the child, and affecting its nutrition, viability, through its mother, prior to birth'.[59] He argued for this approach because 'well-meaning philanthropists so commonly assume that infant mortality is due almost entirely to imperfect or improper feeding of the child'.[60] From his perspective, however, it was impossible to dissociate the problem from its ante-natal causes. He concluded his remarks by including the unborn in the community. We should, he noted, 'realize the inherent sagacity of Burke's definition of a community, viz., that it is "a partnership not only between the those who are living, but between those who are living and those who are dead, and those who are to be born" '.[61]

---

[55] Ibid. 152.
[56] Idem, *Expectant motherhood*, London 1914, 249.
[57] Ibid. 239.
[58] Ibid.
[59] Dr Cooper Patten, 'The ante-natal causes of infantile mortality', *PH* xxiii (1910), 330.
[60] Ibid.
[61] Ibid.

Each of these writings suggests a transition in the discussion of infant mortality. Medical men were really beginning to focus on the conditions which affected foetuses rather than, as in the 1870s, those that often killed infants within the first year or so after their birth. In each instance, the subject of women's work in the dangerous trades was specifically singled out for special attention. Such labour was considered a particularly significant problem which required more attention and even more regulations. Patten's idea of the inclusion of the unborn in the community also suggests the beginning of a new status for the unborn. In this respect, the foetus was beginning to acquire rights that needed to be protected.

Lead is a racial poison

The preceding works suggest an overriding concern with the unborn child rather than with the women workers. This concern was also reflected in other medical writings that linked women's work in dangerous trades to the future of the race. In 1906, for instance, the MOH for the pottery district of Longton said that until there was legislation preventing women from going out to work there would be 'a regular slaughter of innocents in Longton'.[62] In 1911 Dr W. F. Dearden discussed the subject of child and female labour with explicit reference to the fact that 'the future of the race is now regarded of great national importance'.[63] After noting the negative implications of child labour, he said:

> It is also recognized that another result of the factory system, the employment of female labour, is of similiar interest in the same direction [the future of the race is of great national importance], one particular feature, of course, being employment during childbearing and immediately afterwards. Further, it is well known that women and juveniles are particularly injured by laborious processes and in certain poisonous trades.[64]

Because of the prevalent racial concerns, he called for greater restrictions on women's employment during childbearing.

In the end, it was Dr Oliver who played the leading role in highlighting the racial ramifications of women's work. His most dramatic representation of the threat of lead poisoning to the race was made in 1914 when he labelled lead 'a racial poison' which 'destroys the foetus in utero directly or it cuts short its stay in the womb by its action upon uterine muscular fibre'.[65] This

---

[62] *Report of the chief inspector of factories for 1906*, 234.
[63] Dr W. F. Dearden, 'The relation of public health to industrial diseases', *PH* xxiv (1911), 209–10.
[64] Ibid. 210.
[65] Dr Thomas Oliver, *Lead poisoning from the industrial, medical, and social point of view*, London 1914, 184.

was a theme that he had been developing from at least 1902. He did so, for example, during a discussion of the value of the medical inspection of workers in dangerous trades at the British Medical Association's annual meeting in 1902. Oliver was one of numerous participants who praised the beneficial effects of examining women in lead works. In answering the question of who benefits from this state-mandated procedure, he answered 'Personally, I think both the workers and the employers; also, as far as the white lead workers are concerned the race generally, for lead has a prejudicial influence upon the progeny which must not be lost sight of.'[66] He developed this position further the following year when he presided over the Conference on Industrial Hygiene. The theme of his presidential address was whether or not such work was hastening the degeneration of the race. He told his audience:

> There are certain trades that are more harmful to women than to men in consequence of the injurious influence they exercise upon the childbearing powers of women.... Women if pregnant can scarcely follow their work in a lead factory without miscarrying; if they proceed to term, the children are born dead or they die shortly after birth from convulsions.[67]

He claimed that it was this waste of infantile life and 'knowledge of the greater susceptibility of the female to plumbism' that led him to propose their elimination from the dangerous processes in that trade. After surveying the general conditions of women's labour he concluded that 'If women's labour in factories, etc., is thus becoming more prejudicial to their own health, and through them upon their children, then obviously the increasing employment of women cannot be helpful to the race.'[68] He ended his remarks on the subject by wondering if the increased employment of women should be regarded as 'a sign of advancing civilisation, or as a return to the primitive relations of an age long since past'.[69]

Oliver's belief that women's employment outside the home was impeding the advancement of civilisation was clearly expressed in a lecture delivered to the Eugenics Education Society in 1911. 'From the purely physical point of view', he told his audience, 'the nation's strength is measured by its reproductive powers and the high percentage of the fitness of its children.... Women's work becomes the cause of physical degeneracy and of the inability on the part of women to rise to the dignity of the completed act of motherhood.'[70] For that very reason, he had tried to 'emancipate' women from the white lead trade thereby saving the future of the race from the devastation of

---

[66] 'Proceedings of the section of industrial hygiene and diseases of occupation', *BMJ*, 13 Sept. 1902, 745.
[67] Dr Thomas Oliver, 'Address to the conference on industrial hygiene', *JSI* xxiv (1904), 180.
[68] Ibid. 183.
[69] Ibid.
[70] Idem, 'Lead poisoning and the race', 1096.

lead poisoning. After quoting statistics of infant mortality in the pottery trade, he noted, 'The facts which I have just related are of interest from the national point of view, since they deal with an enormous loss of infantile life from immaturity and still-births.'[71] While acknowledging the immediate benefits of state intervention in those trades where lead affects potential motherhood, he concluded that their permanent effect would only become clear in two or three generations.

Medical men, sexual science and lead poisoning

'Occupational diseases and disabilities', Paul Weindling has written, 'provide a sensitive index of social conditions in industrial societies.'[72] The medical diagnosis of the lead-poisoning problem does provide insight into the intertwining of socially constructed ideas about sexual difference with medical theories. The gendered assessment of the lead-poisoning problem, I suggest, was influenced by the development of sexual science, transformations within the medical community and the imperial political climate. I will confine my remarks to the first two subjects; the last one will be developed more fully in the epilogue.

Sometime in the late eighteenth century, Thomas Laqueur and Londa Schiebinger have argued, sexual difference was discovered. There was the shift from the 'one sex' to 'two sex' model which emphasised anatomical differences between the sexes.[73] Instead of viewing women as simply an inferior version of man, they were now seen as entirely different. Their reproductive organs, accentuated in anatomical drawings, became their dominant feature. This 'discovery' of biological difference, Laqueur maintained, became the foundation for a whole series of other differences – cultural, social, intellectual and political – between men and women. The multiple differences were neatly encapsulated in what has been termed separate sphere ideology which placed men within the public world of work and politics while women were relegated to the private world of home and family. Over time, women's nature was progressively constricted as the moral world of the early nineteenth century gave way to the scientific age. Science provided more certainty about sexual difference and 'natural' social roles for the different sexes. Biological determinism replaced moral approbation as women 'evolved' from productive, sexual beings to essentially asexual, reproductive ones. This emphasis on the centrality of women's reproductive organs was clearly seen in gynaecological writings of the nineteenth century. As Ornella

---

[71] Ibid. 1097.
[72] Paul Weindling, 'Linking self-help and medical science: the social history of occupational health', in Weindling, *The social history of occupational health*, 2.
[73] See Laqueur, *Making sex*, ch. v, and Schiebinger, *The mind has no sex?*, ch. vii.

Moscucci has written 'With the redefinition of the ovaries as autonomous control centres of sex and reproduction in the female, it rapidly became a virtually undisputed tenet of gynecological theory that the ovaries, . . . were the essential difference from which all others flowed.'[74]

The development of evolutionary theory during the late nineteenth century provided further 'scientific' evidence to buttress such pronouncements. And it emerged at a critical time, when separate sphere ideology was under assault in a variety of ways. Women were seeking access to higher education and politics, via the suffrage, and entering into previously male occupations and professions. These disastrous developments could be fought with 'facts' from the latest developments in scientific knowledge. For instance, Herbert Spencer argued that women's 'somewhat-earlier arrest of individual development' enabled them to conserve energy for their primary task of reproduction.[75] Patrick Geddes and J. Arthur Thompson claimed that their 1899 publication, *The evolution of sex*, provided the definitive theory of sexual difference. They were convinced that differences in male and female cell metabolism determined all other differences, including social differences. In one of their most famous pronouncements they asserted that sex roles had been decided in the lowest forms of life and 'what had been decided among the prehistoric Protozoa cannot be annulled by Act of Parliament'.[76] Woman's place was in the home and their presence in the labour force was evil because it violated the so-called natural, biological order.

Many historians of gender and science have concluded that such theories about the immutable natural differences between the sexes were infused with the prejudices of their creators and reinforced the prevalent separate sphere ideology. Writing about Victorian biologists, Susan Sleeth Mosedale has argued that their views about women rendered their thinking 'a sterile exercise' and that the exercise itself was 'most frequently taken *for the purpose* [her emphasis] of lending scientific authority to one side of a social issue' to which they were already committed.[77] Janet Oppenheim has contended that evolutionary biology 'provided novel and effective means of justifying old prejudices' and that medical practitioners were 'not reluctant to incorporate those arguments in their work'.[78] The new knowledge, then, enabled them to provide additional proof that the inequality between the sexes was natural and immutable. Such opinions were hard to refute because of the rising status of science and had major implications for women.

Carol Dyhouse has illustrated the specific impact of the new ideas upon

---

[74] Ornella Moscucci, *The science of woman: gynaecology and gender in England, 1800–1929*, Cambridge 1993 edn, 34.
[75] Herbert Spencer, cited in Conway, 'Stereotypes of femininity', 141.
[76] Geddes and Thompson, cited in Easlea, *Science and sexual oppression*, 147.
[77] Mosedale, 'Science corrupted', 54.
[78] Janet Oppenheim, *Shattered nerves: doctors, patients, and depression in Victorian England*, Oxford 1991, 186.

married women workers. Her work on the pre-war medical diagnosis of the infant mortality problem illustrates that many medical men believed that it was caused by married women's labour despite increasing information on circumstantial factors, such as poverty and unhealthy living conditions, contributing to it. She attributed this analysis to the fact that many medical men espoused separate sphere ideology. The writings of Dr Arthur Newsholme, medical officer to the Local Government Board, clearly exemplify the intermixing of such ideas with medical theory. For he refused to abandon the conviction that the employment of women led to a significant loss of infant life, Dyhouse has written, even though he was unable to marshall convincing evidence to support his theory. When, for example, his first report on infant mortality, dating to 1909–10, did not enable him to make any precise statement about the influence of the one upon the other, he still insisted that 'Such employment must, however, tend on the balance to increase infant mortality and to lower the health of older children in the same family. Even when the mother's earnings are necessary for the bread-winning of the family such earnings are secured by some sacrifice of the interests of the next generation.'[79] His objection to the female breadwinner was made more clearly when he commented in his *Report on infant mortality in Lancashire* that 'In a wider sense all industrial occupation of women whether married or unmarried, may be regarded as to some extent inimical to home-making and child care.'[80] They were, of course, her 'natural' duties.

I would suggest that Dr Oliver's thought followed a similar trajectory. His opposition to women's work was made abundantly clear in his remarks regarding the pottery and white lead trades. And he did not confine his remarks about dangerous work to just those two lead trades; other remarks suggest that something other than professional concern for at-risk women was involved. Consider, for example, his contribution to the discussion on enacting regulations for the hand file cutting trade of Sheffield.[81] The state's proposal to limit the hours of work in domestic workshops quickly led to a discussion of women's employment because women comprised the majority of the 2–300 file cutters working in them. Very significantly, there was no discussion of the greater susceptibility of these female workers to lead poisoning. But, then again, no reason existed for such an assertion: men outnumbered women in treatment for the disease at the Sheffield Infirmary.[82] There were no horrific statistics of infant mortality to rely on either. But, according to

---

[79] Carol Dyhouse, 'Working-class mothers and infant mortality in England, 1895–1914', *Journal of Social History* xii (1978), 93.
[80] Ibid.
[81] It was one of twenty trades investigated by the government's 1896 Dangerous Trades Committee which included Dr Oliver. The Home Secretary declared it a dangerous trade in 1902, draft regulations were issued and a further investigation was held in 1903.
[82] For example, out of thirty-four cases of lead poisoning treated at the Sheffield Infirmary in 1894–5, thirty were men and only four women.

Oliver, there was still a problem because when women worked 'Domestic and family duties come to be disregarded by the mother, for she, no less, than the other members of the family, interruptedly lends a hand to increase the income of the home.'[83] In other instances, Oliver generalised about the difficulties inherent in all women's work. 'Say what we may', he told his audience at the Sanitary Institute in 1903,

> the problem of women's labour in all industries is very largely one of sex. Apart from physiological conditions that periodically interfere with women during 25 to 30 years of her adult life, marriage of itself often prevents her from following her usual occupation. The claims too of expected and realised maternity cannot be ignored.[84]

This statement reflected his assumption that all women would marry and become mothers. He added that paid employment outside the home 'takes the women away from their natural duties which should be in the home, and unfits them for becoming mothers'.[85] He then catalogued a chain of disastrous effects set in motion by women's work, including its effect on proper eating habits. Indulging in, he said, 'expensive and savoury articles of her diet, her children acquire, at far too early an age, the same morbid appetite for artificial food as herself'.[86] This last comment echoes some of those made by critics of women's employment outside the home in the 1870s.

The gendered nature of scientific theory was a crucial dimension to the discussion of dangerous work for women but not the only one. There was, as Frank Mort has persuasively shown, an important transformation in medical analysis of the development of diseases. The acceptance of germ theory as well as Pasteur and Koch's investigations into viruses and vaccines led to the belief that disease originated at the cellular level. This perspective led to a shift from earlier nineteenth-century environmental explanations for diseases (and social problems) to bodily explanations.[87] Since the body, rather than the environment, was the site of health or disease it became the locus of investigation. This, in turn, affected public policy initiatives in the decades prior to the First World War when doctors provided medical scrutiny and surveillance of the bodies of key individuals within the population.

The maternal body was singled out for careful monitoring in lead trades because medical men viewed lead poisoning as a feminine problem. Their sexual tunnel vision, as it may be called, was demonstrated by the fact that they were aware of the harmful impact of lead on male virility and hence the male role in producing dead or unhealthy children yet did not make an issue of it. In his 1891 Gaulstonian lectures, Oliver devoted a single sentence to

---

[83] Oliver, *Dangerous trades*, 344.
[84] Idem, 'Address to the conference on industrial hygiene', 181.
[85] Ibid. 182.
[86] Ibid. 183.
[87] Mort, *Dangerous sexualities*, 163.

Table 5
Lead poisoning in the white lead trade, by sex, 1900–14

	1900	Average 1903–5	Average 1906–8	Average 1909–11	Average 1912–14
Male	325	100	83	34	24
Female	53	5	3	2	3

Source: Anderson, *Women in the factory*, appendix ii.

the subject after numerous paragraphs on the horrible effect of lead on women's reproductive system. 'The pernicious influence of lead', he said 'affects equally the reproductive organs of the male, and through the spermatozoa cause abortion, or, if the child is born, it is ill-nourished, and dies almost immediately after its birth, generally from convulsions.'[88] I would highlight the fact that he employed the word 'equally' to describe the impact of lead on the male reproductive organs. By 1902 his remarks were more tentative: 'If lead exercises a prejudicial effect upon the reproductive powers of women it is also capable, although to a less degree, of diminishing the virility of men.'[89] In 1914 he noted 'Contrary to my own experience and that of many other physicians, Carozzi and Frogia are of the opinion that the influence upon the progeny of a father who has been employed in lead is even greater than that of an expectant mother who has been similarly exposed.'[90] Likewise his colleague Dr Legge continued to refer to the harmful impact of work in the lead trades on women but claimed that it was more difficult to draw a definitive conclusion about men. Where there were no precautions to protect men, he said, 'the effect on their offspring . . . may well be evident, as was shown several years ago in the manufacture of pottery in Hungary . . . and again in some other industries cited by Dr Alice Hamilton in her book, *Industrial poisons in the United States*'.[91] The fact that doctors paid scant attention to the link between lead work and diminished male virility suggests that male bodies were not as important in the contemporary discussion of reproduction.

Principal Lady Inspector Adelaide Anderson noted that the emphasis on maternity 'tended to obscure any general perception of the highly urgent need to protect men'.[92] She collected data (*see* table 5) illustrating the preponderance of lead poisoning among men in the white lead trade. Of the reported cases twelve were fatal, eleven of them males and one female. There was further evidence that the state should have been more concerned about protecting men from lead poisoning. In a 1912 work Dr Legge and Dr Goadby

---

[88] Oliver, 'Goulstonian lectures', 21 Mar. 1891, 630.
[89] Idem, 'Lead and its compounds', 303.
[90] Idem, *Lead poisoning*, 184.
[91] Legge, *Industrial maladies*, 65.
[92] Anderson, *Women in the factory*, 114.

Table 6
Lead poisoning in the white lead trade, by sex, 1900–9

	Severity of symptoms			
	Severe	Moderate	Slight	Total
Male	317	235	593	1,167
Female	27	11	33	76

Source: Legge and Goadby, Lead poisoning and lead absorption, table iv.

collected and analysed government statistics of lead poisoning in more than eighteen trades for the period 1 January 1900 to 31 December 1909. According to reports from certifying surgeons, five times more male than female workers were ill: 5,637 cases among the former compared to 1,001 among the latter. Likewise 1,588 males compared to 204 females suffered from severe attacks. By severe attacks they meant paralysis, convulsions and mental afflictions; workers' constitutions were also undermined by paralysis, renal diseases and arterio-sclerosis. Table 6 illustrates, once again, the increase in illness among men after women were banned from the dangerous portions of the white lead trade in 1898. And, the disparity in the numbers is quite dramatic: 1,167 males compared to 76 females.

In addition 490 males were reported ill in the pottery trade compared to 572 females (see table 7), with a respective attack rate (averaged per year for 1907–10) of ten and nineteen per thousand. Legge and Goadby noted, moreover, that while the number of severe cases among male potters was below the average, 'the figures for chronic lead poisoning and paralysis are distinctly high'.[93] With the exception of the latter trade, the total number of male and female workers employed in the various trades was not reported so that determining further statistics of proportional illness is not possible. Despite this methodological drawback, it is obvious that lead poisoning was a problem for both male and female workers.

Focusing on women's bodies also meant that environmental factors contributing to the development of this occupational illness were overlooked. This is quite a significant oversight because the various medical writings reflect their authors' awareness of the important impact of the sanitary conditions of the workplace on the development of this industrial malady. For example, in 1912 Legge and Goadby explicitly made the point that lead poisoning was not evenly distributed throughout the pottery trade. In fact, a minority of potteries were 'sick' factories. They noted that

---

[93] Dr Thomas Legge and Dr Kenneth Goadby, Lead poisoning and lead absorption: the symptoms, pathology, and prevention, with special reference to their industrial origin and account of the principal processes involving risk, London 1912, 50.

Table 7
Lead poisoning in the pottery trade, by sex, 1900–9

	Severity of symptoms			
	Severe	Moderate	Slight	Total
Male	102	158	216	490
Female	86	181	286	572

Source: Legge and Goadby, *Lead poisoning and lead absorption*, table iv.

among the 550 potteries, in the years 1904 to 1908, five potteries were responsible for 75 cases, and 173 for the total number of cases, leaving 377 factories from which no cases were reported. . . . Particular factories, owing to special method of manufacture or special manner of working, may have an incidence out of all proportion to that prevailing in the trade generally.[94]

Earlier medical reports had frequently emphasised that the conditions of the workplace were a major factor in the incidence of disease. In 1893 Arlidge had argued that in the majority of cases the development of lead poisoning could be explained by obvious factors: carelessness at work, eating and drinking amidst the lead and dust and the overall sanitary condition of the particular workplace. He further asserted that women were more readily affected by poison and attributed this to their long hair and dress, upon which the poisonous dust would accumulate, and their 'unwillingness to lessen their proclivity with suitable coverings'.[95] Other observers, including Oliver, acknowledged that women's health in lead works was affected by their physical condition prior to work, their diet and the sanitary conditions of their homes. In the end, these recurring points were under-emphasised in the analysis of the lead poisoning problem.

Lead poisoning was a serious problem affecting both female and male workers. It was a complicated problem caused by a variety of factors and yet medical men reduced it to a 'woman's problem'. Social ideology, as well as a preoccupation with the need to protect the maternal body, influenced the development of the theory that sex determined susceptibility to lead poisoning and that lead poisoning was a special problem for women. Moreover, prevailing views about masculinity, that men were independent, self-reliant workers, impeded the development of necessary male protection. This interpretation, newly articulated by Oliver in the 1890s, became the reference point of the subsequent discussions of lead poisoning in Britain and America well into the 1920s. For instance, Dr Hope, Dr Hanna and Dr Stallybass incorporated this interpretation into their 1923 historical survey

[94] Ibid. 54.
[95] Arlidge, *The pottery manufacture*, 16.

entitled *Industrial hygiene and medicine*.[96] Meanwhile Dr Edgar Collis and Major Greenwood, who had served as head of the medical branch of the Ministry of Munitions during First World War, were among its critics. Thus they wrote:

> The statement that woman is less resistant to men to environmental conditions and toxic influences has been made so often that it has passed into occupational literature almost unchallenged.[97]

While they believed that lead exercised an injurious impact on unborn children, they 'did not feel satisfied that there is evidence that they [women] fall victims more readily than men'.[98] They based this conclusion upon the fact that during the war women were temporarily allowed to return to work in the white lead beds without any significant rise in lead-poisoning rates. They attributed this development to the efficacy of precautionary measures and the 'resistant powers of women'.[99] Thus they argued that there was no 'scientific evidence in support of the alleged sexual proclivity' of women to lead poisoning.

The activities and writings of Dr Alice Hamilton, the foremost figure in the study of lead poisoning in early twentieth-century America, reflected the influence of Oliver's theory across the Atlantic. She extensively quoted him, as well as Legge, Goadby, Arlidge and Prendergast, in her 1925 work *Industrial poisons in the United States*.[100] She argued that several American investigations of lead poisoning in potteries supported the contention that women were more liable to develop the disease than men. Most significantly, she relied extensively upon Oliver and other British doctors in her chapter 'Lead as a race poison'. 'Lead is often spoken of as a race poison', she began, 'in that its effects are not confined to the men and women who are exposed to it in the course of their work, but are passed on to their offspring.'[101] And, like her British counterparts, she supported the elimination of women workers who came in contact with lead because of its impact on their potential offspring.[102]

As will be seen in the following chapter, various feminist organisations were involved in the regulation of dangerous trades. While some accepted the medical definition of danger others would challenge it and its creators.

---

96 Dr E. W. Hope, Dr W. Hanna and Dr C. O. Stallybrass, *Industrial hygiene and medicine*, London 1923.
97 Dr Edgar L. Collis and Major Greenwood, *The health of the industrial worker*, London 1921, 231.
98 Ibid. 233.
99 Ibid.
100 Dr Alice Hamilton, *Industrial poisons in the United States*, New York 1925, 8.
101 Ibid. 110.
102 For more on Alice Hamilton and the American experience in regulating dangerous trades see Allison L. Hepler, *Women in labor: mothers, medicine, and occupational health in the United States, 1890–1980*, Columbus 2000.

7

Feminists, Dangerous Trades and the State

Did sexual difference mandate dangerous trades regulations? Were they and other regulations necessary because of women's ability to bear children? Was reproduction a private matter or one of public and national interest? Feminist organisations debated the answers to these critical questions in the decades before the First World War. Their divergent positions on dangerous trades were made abundantly clear as they petitioned the government, led deputations to the Home Office and published pamphlets and books. Members of the WIDC and the SPW, under the leadership of Boucherett, characterised dangerous trades measures as a violation of a woman's individual rights. 'The most absolute monarch never was invested with such authority', she and Helen Blackburn wrote, regarding the expansion of the Home Secretary's power in 1895.[1] They especially objected to the fact that working women were not consulted when he and other government officials decided upon the future of their livelihoods. Meanwhile, members of the WTUL, the WIC and the WCG heartily endorsed the extension of state power to protect vulnerable women. They were convinced that legislation was especially necessary when the conditions of women's labour affected their potential offspring. As Tuckwell of the WTUL remarked about the white lead trade, 'the worst feature of this trade was its affect on the next generation'.[2]

These groups had also expressed conflicting views on the broader issue of state intervention and motherhood when the government enacted a mandatory unpaid four-week leave after childbirth as part of the 1891 Factory and Workshop Act.[3] Tuckwell applauded this measure because, she said, it established the principle that the state should and could intervene to protect the health of a woman and her children. Wanting further legislation to solve what she called the problem of married women's work, she advocated in 1894 'the prohibition of the labour of mothers of families until these children shall have arrived at an age to take care of themselves'.[4] The loss of the household's wages from such a prohibition, she believed, 'will be compensated by the fact that her competition in the labour market is also withdrawn, and by the additional thought which she is enabled to give to the ordering and management

---

[1] Boucherett and Blackburn, *The condition of working women*, 21–2.
[2] Tuckwell, *Women's work and factory legislation*, 8.
[3] For a discussion of this provision see Hutchins and Harrison, *A history of factory legislation*, 209–11.
[4] Gertrude Tuckwell, *The state and its children*, London 1894, 161.

of the household when she shall be permanently at its head'.[5] Boucherett and her associates vociferously opposed such state intervention because it violated their cherished principle of a woman's right to determine the conditions of her labour. They asserted that women made their own arrangements and could manage perfectly well on their own. Moreover, they opposed this legislation because of its adverse economic effect upon women and the children whom they may support. 'This [legislation] means', Helen Blackburn and Nora Vynne wrote, 'that the women compelled to work must leave the work that is the best that they could get and take the second best ... the State should support those who cannot support themselves at such a time.'[6] This prohibition, they continued, caused 'grave material evils – evils very much greater than the mere decrease in population likely to follow the penalisation of childbirth'.[7]

New personalities and organisations entered into the dialogue on the hazards of women's work and motherhood after 1900. Margaret MacDonald, Eva Gore-Booth and Esther Roper emerged as leaders of groups primarily concerned with that issue and became key participants in the renewed governmental and public dialogue on dangerous trades between 1903 and 1911. During those years, the government sought to prohibit women from working as barmaids and on the pit brow of coal mines because of the alleged impact of such work upon themselves and, especially, their unborn children. This chapter will analyse these developments to illustrate more fully the complex feminist reaction to the linkage of danger to women's work and the state's regulation of the maternal body. Moreover, it illustrates how phrases like 'dangerous trades' or 'unhealthy trades' had entered into the common vocabulary of pre-war protection. The application of those terms varied considerably from trade to trade, from group to group and very commonly, I argue, they were applied to jobs that were not especially perilous for women but considered unsuitable for them.

Serving working women's best interests

During the enactment of dangerous trades regulations in the 1890s, divergent feminist organisations claimed to represent the true interest of the working women who would be affected by such legislation. This same argument was put forth by newer groups such as the Freedom of Labour Defence Association founded by Boucherett and Blackburn in 1899. Well-known members of the women's rights movement, Heather Bigg and Nora Vynne, joined them to protect 'industrial workers, especially working women from the imposition of legal restrictions which would diminish their wage-earning capacity, limit

---

[5] Ibid.
[6] Helen Blackburn and Nora Vynne, *Women under the Factory Act*, London 1903, 74.
[7] Ibid. 75.

their personal freedom, and inconvenience them in their work'.[8] FOLD pursued the same methods as the SPW and WIDC, with particular emphasis on conducting their own investigations of the conditions of women's work and presenting the working women's point of view.[9] Like their anti-protection predecessors, the group denounced dangerous trades regulations. In 1900 they wrote:

> An impression has been created in the public mind that women are more susceptible than men to lead poisoning, and that they therefore require special intervention on the part of the law. It has seemed to most people remarkable that what is so poisonous for women should be so comparatively harmless to men, so that women had to be turned away, or subject to special restrictions, men might be left to their own devices.[10]

They made repeated references to the lead-poisoning statistics contained in the report of the chief inspector of factories for 1898. The statistics on page 106, they noted, illustrated a greater amount of chronic lead poisoning among men while additional statistics 'shows there is little difference between the two sexes as regards the frequency of colic, anemia, and headaches'.[11] This provided the statistical evidence for their contention that sexual difference did not necessitate special regulations for women. In this, as in every other instance, they tried to negate sexual difference; to portray women as asexual workers because sexual difference had been the basis for their different treatment by the state.

In stark contrast, MacDonald continued to express the social feminist belief that women needed special legislation precisely because of sexual difference. The reason for labour laws for women, she wrote in a 1900 Independent Labour Party pamphlet, 'is simply that men and women are different and so have different needs'.[12] She cited potential motherhood as the key difference between the sexes that gave the state the right to interfere in women's work. Since a woman's actions before and after birth affected her children, she said, 'however much it may interfere with the alleged liberty of working women, it is the undoubted duty and interest of the community to protect, through her, the health of the rising generation'.[13] She claimed that

---

8 FOLD, 'Notes and incidents', *ER* xxxi (1900), 34.
9 They were instrumental in publishing *Statute mongery: their results, the remedy*, London 1901. This title suggests a rebuttal of Margaret MacDonald's tract, *Labour laws for women: their reason and their results*, London 1900. They also published a pamphlet by one of the group's members, Florence Greenwood, who was a sanitary inspector in Sheffield, entitled *Is the high infantile and death rate due to the occupation of women?*, London 1901. Greenwood argued against the idea that women's work outside the home was the cause of infant mortality and explored the other conditions which contributed to it.
10 FOLD, 'Notes of the quarter', *ER* xxxi (1900), 96.
11 Ibid.
12 MacDonald, *Labour laws for women*, 10.
13 Ibid.

the state had a further interest in a woman's life because 'her home duties are not only her personal affair, but concern the community'.[14] MacDonald concluded that the state 'is interested not only that she should bear healthy children, but also that she should have the time to attend to their needs or those of any other members of her family, and so it is specially necessary that she should not have too long hours or too exhausting work'.[15] She was essentially reiterating Tuckwell's argument that protection would enable a woman to fulfil her 'natural' duties of motherhood and household management.

MacDonald served as president of the Women's Labour League, an organisation that included the wives of prominent labour leaders as well as trade unionists Mary MacArthur and Margaret Bondfield.[16] Since this group was centrally concerned with the issue of the industrial employment of expectant mothers, the topic was frequently discussed at its annual meetings. In 1907, for instance, Mrs Bruce Glasier illustrated the dangers of women's work by referring to the 1904 report of the Committee on Physical Deterioration and Dr George Reid's papers on women's work and infant mortality in the Potteries. 'It was perfectly evident', she said, 'that the exploitation of mothers before and after childbirth led to a high death rate among infants.'[17] MacArthur addressed this same issue at the group's annual meeting three years later and criticised the theorists (no doubt a reference to opponents like Boucherett) who failed to realise that certain work was especially dangerous for women and their unborn children. She was careful to differentiate between the types of labour performed by professional, that is middle-class, and working women. A woman, she told her audience, might write a book or paint during pregnancy but it was a different matter if 'she depended on working a heavy machine with treadles for twelve hours a day' or was occupied 'in the dipping house of a Staffordshire pottery. If she was engaged in that industry within a short time of the child's birth, the chances were four to one against the child being born alive'.[18] She and her colleagues in the WLL believed that such conditions of labour warranted special regulations. At the same time, however, they believed that the state should provide financial assistance for women who would be deprived of income around childbirth. 'If we keep a woman at home for race good', Bondfield added, 'the State must see that she is maintained in a state of physical health and efficiency. It is a race question not an individual one.'[19] Since the state had a vested interest in

---

14 Ibid. 12.
15 Ibid. 12–13.
16 For more on the WLL see C. Collette, *For Labour and for women: the Women's Labour League, 1906–1916*, Manchester 1989; C. Rowan, 'Women in the Labour Party', *Feminist Review* xii (1982), 74–91; and Lucy Middleton (ed.), *Women in the Labour movement: the British experience*, London 1977.
17 WLL, *Second annual conference report* (1907), 17.
18 WLL, *Fifth annual conference report* (1910), 11.
19 Ibid. 21.

children it should contribute to their welfare at this critical time in their lives.

Those women were basically restating the well-established arguments for and against protection. The same cannot be said of 'radical suffragists' Gore-Booth and Roper.[20] After the two women, the former an Irish aristocrat and the latter a graduate from Owens College in Manchester, met in Italy in 1896 they began a life-long personal and professional connection.[21] Most notably, they founded the Lancashire and Cheshire Women's Textile and Other Workers' Representation Committee in 1903 and the National Industrial and Professional Women's Suffrage Society in 1906. These groups included both professional and working women who believed that the 'feminisation' of the state, through the vote, was essential for the improvement of women's economic and industrial position. 'Political emancipation must precede industrial emancipation', they proclaimed in the LCWTOWRC's manifesto, because 'the political disabilities of women have done incalculable harm, by cheapening their labour and lowering their position in the industrial world.'[22] If women could vote then they could decide for themselves whether protection was necessary and the form it should take. Until then they opposed protection because, they claimed, its construction was influenced by working women's male competitors and middle-class philanthropists who did not represent the interests of working women. Their idea of 'no protection without representation' was a crucial new dimension to the ongoing discussion of protection and the relationship between women and the state.

Feminists and 'the barmaid question'

An examination of the activities of the different organisations regarding the so-called 'barmaid question' will provide more specific information on feminists' viewpoints about dangerous trades regulations. This question had its origins in an episode in Glasgow in 1902 when licence-holders dismissed all the barmaids in the city, following the magistrates' recommendation made while issuing their licenses.[23] This was followed in 1906 by J. P. Gooch's

---

[20] Jill Liddington and Jill Norris coined the phrase 'radical suffragists' to describe the groups of northern suffragists who had strong links with working-class organisations that flourished in the towns: with, they said, branches of the Women's Cooperative Guild, the Independent Labour Party and local groups of women textile workers. See their pathbreaking work, *One hand tied behind us: the rise of the women's suffrage movement*, London 1978.
[21] For more on their personal and professional life see ibid. 77–88, ch. ix.
[22] Manifesto of the LCWTOWRC, reprinted in Eva Gore-Booth, 'The women's suffrage movement amongst trade unionists', in Brougham Villiers (ed.), *The cause for women's suffrage*, London 1907, 50–62 at p. 51.
[23] For the history of the important Glasgow precedent see the Home Office memo on the 1906 employment of barmaids bill, 12 Dec. 1906, HO 45/10312/124109.

parliamentary bill which sought to stop the employment of barmaids. He proposed the mandatory certification of existing barmaids, whose employment would be terminated at a future date, and the prohibition of new women from entering the trade. After its failure, the government introduced a clause in its 1908 licensing bill proposing to give local magistrates the power to attach any conditions which they thought fit to the employment of barmaids.

The Joint Committee on the Employment of Barmaids, chaired by MacDonald, initiated a debate between feminists over the employment of barmaids. Formed in 1903, this group drew its membership from a wide range of men and women interested in social reform including the bishop of Southwark and MPs Gooch and Arthur Henderson as well as, they carefully noted, members of only three or four temperance organisations.[24] It began its campaign against barmaid employment by sending letters to the licensing sessions in London urging them to follow Glasgow's example and recommend that licence-holders discontinue their employment of barmaids. The JCEB publicly announced its desire for the gradual closure of this trade to women in a letter to the editor of *The Times* in late 1903. In 1904 they outlined the numerous dangers of the trade to women in *The Barmaid Problem*. This pamphlet, sent to the Home Secretary, became the first item in a new barmaid's file in the Home Office. The group's proposals, not surprisingly, were embodied in its member Gooch's 1906 bill.[25]

The organisation graphically outlined the multiple dangers of this trade for women. It began by recounting the evidence of the miserable conditions in bars that had been revealed during Eliza Orme's investigation of this work for the 1892 Royal Commission on Labour. Women uniformly complained to her of long hours and overwork, stating that it was common for barmaids to work between ninety and ninety-five hours per week. The report also disclosed that barmaids were packed into sleeping accommodations that were frequently no more than ill-ventilated closets. Although food was provided with their jobs, its poor quality led many barmaids to purchase their own with their wages. Yet, they argued, while it might be possible to remove some of those dangers there was one which could not be removed: women worked in the 'poisonous air in a drinking-bar, laden with fumes of alcohol and tobacco' which 'irritates the eyes, stimulates thirst, insensibly lowers the system, and leads to debility and anaemia'.[26]

---

[24] This point about the group's membership was made in a letter to the editor of the MG, 30 July 1906, in response to a letter from a Mr Robinson, secretary of the Licensed Victuallers' Defence League.
[25] In November 1904 they sent their pamphlet, *The barmaid problem*, London 1903, along with a letter, to the Home Secretary calling his attention to this problem and expressing the hope that he would support their recommendations should they be included in a bill in parliament: Margaret MacDonald, chairman of the JCEB, to the Home Secretary, HO 45/10312/124109.
[26] JCEB, *The barmaid problem*, 4.

Their account of long hours of work and poisoned air, however, was eclipsed by their description of the horrific moral evils inherent in this work. It was impossible, the JCEB claimed,

> for a barmaid to be in the business for long without gaining knowledge of the life of a fallen women, and without hearing oaths and words, whether addressed to her or forming the theme of conversation among the customers, which will cause her ears to tingle, and will either provoke indignant disgust, which she dare not express, or will tend to the insidious weakening of the barrier of modesty and maidenly shame in which her strength resides.[27]

Bad language and tingling ears were only the beginning of the occupational hazards. Attractive young women, they continued, were hired in order to obtain good pecuniary results. 'Barmaids are sirens who lead young men to drink', wrote one observer, 'and the question is whether the purveyors of alcohol should be allowed to use up such a mass of maidenhood as is annually sacrificed to the trade, merely for the sake of giving additional attractiveness to the drink they sell.'[28] Such young women were then exposed to the unhealthy tone in bars and vulnerable, they claimed, to temptation, seduction and abandonment. The group particularly emphasised that alcoholism was widespread among barmaids and attributed its development to 'the natural result on a women's frame of late hours, excitement, and physical and nervous overstrain, coupled with the temptation of having stimulants always at hand'.[29] Moreover, they concluded, the lives of many of these women ended in suicide because they could not find employment after their barmaid career ended, typically, at the age of thirty-five. In the end, they created a sad and melodramatic account of the moral perils of this trade.

Finally, they included medical testimony as to the hazardous nature of work at the bar and its impact on the future of the race. They reported that Dr Kelynack, superintendent of Northwood Sanatorium, said that 'medical men oppose the employment of women in bars on two grounds – that of the good of the individual, and the benefit of the race'.[30] The future of the race, he argued, was affected when the moral and/or physical effects of woman's labour interfered with the fulfilment of her functions as a mother. According to *The Lancet*, the vocation 'is attended by perils – long hours and frequently excessive fatigue are only too likely to lead to alcoholic indulgences'.[31] Furthermore, this leading medical journal claimed that 'learning the bar', as it was called, had no tendency to make a woman a better wife or mother but rather it 'permanently injures her health, and [. . .] exposed her to exceptional risks of seduction and intemperance, with the consequence both to herself and her

---

[27] Ibid. 5.
[28] JCEB, *Women as barmaids*, London 1905, 17.
[29] Ibid. 23–4.
[30] Ibid. 16.
[31] Ibid. 18.

offspring that these conditions entail'.[32] Was it consistent, they asked, with due regard to the national welfare, to 'allow the daughters of the Empire to be offered up as sacrifices to the Moloch of the drink trade?'[33] According to the *Medical Press* many barmaids 'lapse into intemperance, and from the conditions of work, morbid psychical and abnormal physical states are readily induced. Speaking purely from the medical standpoint there can be no hesitation in declaring that the State should protect itself by safeguarding its women from such dangerous conditions of life as are at present inseparable from the majority of drinking-bars'.[34] Finally, they noted that Dr Fleck, medical superintendent of the Bentry Inebriate Asylum, also connected this problem to future motherhood: 'If we could allow the young girl to have a chance of better fitting herself for becoming a wife and mother, without the temptations to this most disastrous calling, we should reduce intemperance amongst women.'[35] Because of this testimony, they contended that the employment of women in bars was obviously dangerous for the future of the race.

The JCEB urged the government to close this lurid and dangerous trade to women because of its moral, social and physical dangers. They cited the prohibition of women from working underground in the mines in 1842 and from certain processes in the white lead trade in 1898 as precedents for their rather drastic proposal. 'The latter prohibition', they said, 'was enacted solely for reasons of health; but in the crusade to stop women and girls from working underground in mines, morals as well as sanitary conditions had great weight.'[36] To buttress their position they noted that their campaign had the support of important organisations such as the National Union of Women Workers and the WLF. They cited the former group's 1903 condemnation of bar work as undesirable for women as well as the WLF's 1904 resolution calling for the prohibition of the further hiring of women in bars. Anticipating criticism, they argued that the number of women who would be displaced was relatively small and that they would be able to find alternative employment.

The response to this work was, as expected, mixed. The WIC characterised their study of barmaids as excellent. In 1906 they claimed that

> We own to having a theoretical prejudice in favour of regulation rather than the prohibition of women's work except in extreme cases, such as underground work in mines, but it seems to us that this report has shown good grounds for regarding the barmaid's work as one of those extreme cases that do justify extreme remedy.[37]

[32] Ibid.
[33] Ibid.
[34] Ibid.
[35] Ibid.
[36] Ibid. 57.
[37] WIC, 'The employment of barmaids', *WIN* xxxiv (1906), 542.

FOLD, on the other hand, claimed that opposition to barmaid's work was based upon exaggerated statements that were not representative of most women's experiences. Most important, they raised the persistent question of whether or not the government had the right to close off a trade that provided women with the opportunity to earn an honest and respectable living.[38] Most significantly, Gore-Booth and her associates strenuously disagreed with the group's assessment of the dangers of the trade and especially their proposed solution to the 'barmaid question'. As a result, she became the leader of a countermovement that involved a campaign in the press, the foundation of the Barmaid's Political Defence League in 1907 and several deputations to the Liberal Home Secretary, Herbert Gladstone.

Not surprisingly, Gore-Booth and members of the LCWTOWRC initially expressed their opposition to the JCEB in the press – which actively covered this story and provided publicity for their efforts. Consequently, the opening salvos between Gore-Booth and MacDonald occurred in the editorial pages of the *Manchester Guardian* in late 1903 and early 1904. Gore-Booth and Sarah Dickensen's first letter attacked the group's activities stating that 'To anyone acquainted with the present state of the labour market and the difficulty of obtaining employment it seems a policy dictated by the most *short-sighted philanthropy.*'[39] Given those circumstances, they questioned how her group could recommend the prohibition of this honest work for women while others were struggling to find jobs. MacDonald, in turn, critiqued their outdated attitude that 'laissez-faire must reign supreme' and made a general defence of protection. 'Special regulation of hours and of sanitary conditions for women', she noted, 'have not handicapped them in their work in factories and workshops, but immensely improved their position as wage-earners.'[40] While acknowledging the seriousness of closing an entire trade to women, she believed that it was necessary in this particular case because, she wrote, 'we consider the circumstances of work so disadvantageous'.[41] Gore-Booth and Roper totally disagreed and remarked that

> We fail to see any parallel between the Factory Laws as observed by the women in Lancashire. . . . Regulation which has for its aim the amelioration of the hours and conditions of work, sanitary and others, is not to be confused with a totally different issue, the proposal by a gradual process of weeding out to supplant one set of workers with another.[42]

This interchange outlines Gore-Booth and MacDonald's arguments, arguments that would be elaborated in 1907 and 1908.

---

[38] FOLD, 'Barmaids', ER xxxiv (1904), 23.
[39] MG, 23 Dec. 1903.
[40] Ibid. 16 Jan. 1904.
[41] Ibid.
[42] Ibid. 4 Feb. 1904.

'The worst of its [the government's] encounters with the Women's Suffrage Party is yet to come, and the reforming zeal of some of its members, in proposing to abolish barmaids, has brought a fresh group of angry women clamouring round the doors of Cabinet Ministers', the *Daily Telegraph* proclaimed in its article entitled 'Disbarring barmaids' that reported on the first barmaids' protest meeting in London in March 1907.[43] While MacDonald lobbied parliament to eliminate this field of employment because of its alleged moral and physical dangers to women, Gore-Booth, Roper, Dickensen and Sarah Reddish organised women in the trade and mounted a campaign to fight this 'shortsighted philanthropy'. This meeting, packed with barmaids and their female and male sympathisers, voiced their protests, appointed a deputation to meet the Home Secretary and led to the formation of the Barmaids' Political Defence League.[44]

Speakers denounced the JCEB for undue interference in the lives of these working women whose labour and life they misrepresented in their publication, *Women as barmaids*. Those pressing for the abolition of barmaids, according to Gore-Booth, were entirely unfamiliar with the economic conditions of women's labour. They seemed to think, she said, that they could turn women out of their jobs for their own good without inflicting hardship upon them. Miss Kate King-May, a FOLD member active in this campaign, accused them of being middle-class women who had never done a day's work in their lives yet believed they were 'heaven-sent directors of other people's morals and destiny'.[45] They then deconstructed the information presented in the group's publication to prove that they had levelled untrue allegations against barmaids and their profession. For instance, they seized upon the contention that barmaids were addicted to drink and led irregular lives. Yet, they noted, *Women as barmaids* only recorded eighteen cases of drunkenness and sixteen cases of irregular living out of 27,000 women currently employed as barmaids. Dickenson responded to the negative health comments by saying that 'Judging by those [barmaids] she saw in the meeting, they were just as good-looking and healthy a class of women as could be found in any other trade.'[46] The last remark elicited cheers from the enthusiastic and supportive crowd.

Roper connected this attack on women's employment to their disenfranchisement. The cause of all this trouble, she said, was a government which had no sense of responsibility towards women. If they had to ask women for votes, she exclaimed, there would be a lot less talk of passing such legislation directly affecting them. The women at the meeting enthusiastically agreed

---

[43] *DT*, 15 Mar. 1907.
[44] Eva Gore-Booth outlined the group's objectives in *The Barmaids' Political Defence League*, Manchester [1907/8].
[45] *DT*, 15 Mar. 1907.
[46] Ibid.

and unanimously passed a resolution to be sent to the Home Secretary stating that

> They deeply resent the proposal of the Home Secretary to give the authority of the State to that section of the public, who, in their efforts after strict temperance legislation, do not shrink from swelling the numbers of the unemployed workers, increasing the sharpness of competition, and the sum of poverty in the country.[47]

This legislation was itself dangerous, Roper exclaimed, for although it did not directly propose to deprive barmaids of their present employment 'it was the thin edge of the wedge, and if it were passed it would not be long before another measure would be passed empowering magistrates to say that barmaids should no longer be employed on licensed premises'.[48] They then selected, from a mass of volunteers, a deputation to meet the Home Secretary.

The BPDL expressed its anger and resentment against the proposals at deputations to Home Secretary Gladstone in both 1907 and 1908. Nine barmaids, one of whom had worked forty years in the trade, were members of the first one. As a group, they protested at being publicly maligned and insulted. 'People', Gore-Booth said, 'talked of temptation to immorality, but they ought to realise that the worst temptation to immorality was a state of unemployment, absence of wages, and starvation.'[49] Moreover, they claimed that about 100,000 women would be affected by the proposals emanating from ignorant reformers. Perhaps, she said, they did not realise that closing a trade that absorbed so many thousands of women would throw them into the overcrowded labour market. In view of the present unemployment crisis, they believed that was an especially grave mistake.

In addition to that familiar line of argument, the group also tried to defend the profession of barmaid by comparing it favourably with other work. Dickensen, for example, told Gladstone that she had worked in a licensed house and a cotton mill and did not see why the barmaid trade had been singled out as especially harmful for women. 'Her experience', she said, 'convinced her that, as far as hours and wages were concerned, they were practically a great deal better off than some of the women in industrial occupations.'[50] Roper added that the average wage of a barmaid was 12s. a week with board, lodging and washing, a sum comparable to wages earned by shop assistants. Barmaids usually worked twelve hours a day, comparable to domestic service, but they could rest between customers.

Gladstone assured them that he 'for one had never joined in any condemnation of the barmaids as a class, or in any imputations against them'.[51] More-

[47] Ibid.
[48] Ibid.
[49] MG, 29 Nov. 1907.
[50] Ibid.
[51] Ibid.

over, the government realised 'that any drastic provisions to check or diminish or put an end to the employment of women on licensed premises might have a result far more immediately disastrous than any mischief which can be shown under the present system'.[52] They would carefully consider any action that would interfere with the employment of women but still wanted to enact some legislation which would produce proper conditions for the minority of licensing houses in which women were subjected to poor conditions. The Home Office, which had anticipated 'the usual opposition to any interference with women's employment', responded to the BPDL's concerns in a sympathetic manner. A Home Office memo on the 1906 proposals had noted that

> The profession of Bar attendant may be claimed as one of the occupations in which a women is on an equality with a man, if not at a considerable advantage over him, and the Bill proposed to eliminate her competition altogether. The Bill may even be denounced as the thin edge of a wedge which the thicker would be perhaps be the exclusion of women from unlicensed refreshment rooms – Tobacconists and Sweet Shops and even Post Offices. Indeed the reasons which are said to have actuated the Glasgow magistrates who held that the profession of Barmaid was undignified, unbecoming, and degrading to a woman, give countenance to this view.[53]

Whitelegge also commented that neither of the precedents, the underground mine and the white lead bans, cited by MacDonald's group 'are very close precedents although it is said that underground work in mines was prohibited for moral as much as sanitary grounds'.[54]

Such Home Office reservations, combined with a strong publican lobby, helped defeat the bill as it passed through parliament so that the barmaid trade remained open to women. Despite this victory, Gore-Booth and Roper were angered that they had had to defend a woman's right to work against other women's groups. They also knew better than most about the larger implications of the barmaid campaign because it was one of several campaigns involving their groups. Two years before the barmaid victory, they had organised women acrobats and defeated the 1906 Dangerous Performance (Women) Bill, slated to abolish them because acrobatics was considered especially unseemly and dangerous for women. It turned out that women were not the performers involved in a series of accidents that had prompted government concern for this type of work. In 1909 they mobilised florists and their assistants after the government rescinded the special exemption (from the existing factory and workshops regulation) that had allowed them to work in the evening. The WTUL, the WIC and Labour MPs had persuaded the government to take this course of action because they objected to the

---

[52] Ibid.
[53] Home Office memo, 12 Dec. 1906, HO 45/10312/124109.
[54] Whitelegge minute, 27 Feb. 1907, ibid.

irregular hours of work for women employed in this trade. Gore-Booth and Roper argued that this measure would drive employers to replace these skilled and well-paid women workers with foreign men.[55] Much to Gore-Booth's dismay, the government stood by its decision to rescind the special exemption.

'They walked like Juno': pit-brow lasses

Ever since women had been banned from working underground in mines in 1842, they had found work on the pit heads pushing tubs of coal to a belt where other women sorted it. In the decades before the First World War, these 'pit-brow lasses' joined barmaids, as well as female nail and chain, white lead and pottery workers, as subjects of intensive public scrutiny and controversy. Legislative proposals to eliminate them from pit-brow work were put forward in 1886–7 and 1911. Angela V. John has extensively analysed the earlier episode so I will concentrate on the later one.[56] On 1 August 1911 Sir Arthur Markham, Liberal MP for Mansfield and colliery owner, proposed an amendment to the 1910 coal mines bill stating that women who had begun pit-brow work that year should not be allowed to work after the first of the next year and that no more women be engaged after that date. He made that proposal, he said, to end the endangerment of women, especially married women, who had to push tubs and perform generally strenuous work. His proposal immediately sparked a controversy over the fate of some 5,000 women from the Wigan area in the North of England who might possibly lose their jobs.

Not surprisingly, Gore-Booth and Roper organised meetings in Wigan, Manchester and London to protest at this legislative interference in women's work.[57] At a crowded September meeting in the Albert Hall in Manchester, they passed a resolution that 'Their work was neither too rough nor too heavy, alertness and skill being the qualities most necessary for the worker. They have already an eight-hours day and the conditions of open-air work are healthy.'[58] At the London meeting in October, Gore-Booth called this legislation an attempt 'to manufacture artificial unemployment' and remarked that if women did not get the franchise 'twenty-five years hence we shall be saying the same thing. But if you can turn MPs out, they are not so keen to

---

[55] Gore-Booth expressed her vehement opposition to this course of action, for example in 'No trades for women', *Englishwoman* i (1909), 507–17, 'The Home Office and the florists', CC i (1909), 501–2, and a letter to the editor, MG, 14 July 1909.

[56] See Angela V. John, *By the sweat of their brow: women workers at Victorian coal mines*, London 1984, chs v–vi.

[57] Their first meeting was held in Manchester on 18 Sept. 1911, followed by meetings in Wigan and London on 7 and 31 Oct. respectively.

[58] Circular and letter from the LCWTOWRC and NIPWSS to Home Secretary Herbert Gladstone, 17 Oct. 1911, PRO, POWE 8 H/S 164/206.

turn you out'.[59] At both meetings speakers reiterated the now familiar demand for the suffrage as a defence against any further legislative restrictions on women's work.

The LCWTOWRC sent the Home Office its pamphlet, *Statement of an amateur pitbrow worker* which, Gore-Booth claimed, presented irrefutable evidence that this trade was not dangerous to a woman's health. The statement came from King-May, a lecturer in physical training for female students enrolled in the department of education at Manchester University, who had previously been involved in the defence of the barmaids. After reading about the controversy over pit-brow work she travelled to Wigan and obtained a job at the Park Lane Colliery and her experiences formed the basis of this literature. First, she attested to the fact that the lives of pit-brow workers were neither rough nor uncivilised. Living among them, King-May found them to be 'self-respecting, thriving working people. The women were good looking and well built, hard workers, cheerful and pleasant people'.[60] Their wages were poor compared to men's wages but at 10s.–12s. a week they were on par with women's wages in other trades. She worked, between 6.30 a.m. and 6 p.m., at several jobs but primarily at 'tub shoving' and sorting the coal. Tub shoving, she suggested, was a misleading expression for the job since it involved nothing more than pushing a tub with one hand, if one were a delicate women, with two fingers if one were a strong woman. She was able, she said, to push the tubs along the steel plates to the weighing machine with a sharp pull of two fingers. Nor, she remarked, was coal sorting a difficult job. She described her co-workers as healthy looking and added 'when so much work is done now-a-days under bad conditions, ill-ventilated, and unhealthy rooms, from a medical point of view it seems all important that this work should be kept open for women who are peculiarly likely to profit from its healthy out-of-door nature, and good exercise'.[61]

Unlike the barmaid controversy, this campaign to eliminate women's work was universally condemned by diverse women's organisations. This included other suffrage groups such as the WCG which claimed that this episode was 'another object-lesson for the necessity of women having the Parliamentary vote'.[62] Members of the National Union of Women's Suffrage Societies, the umbrella organisation led by Millicent Garrett Fawcett, published an account of their meeting with a group of pit-brow workers in their paper the *Common Cause*. They reported that the women presented themselves

> with perfect self-possession and the good manners that come from being right. They were emphatically 'right,' and it makes one rage to think that some men

---

[59] A report of this meeting was published in National Union of Women's Suffrage Societies, 'The pit-brow women's protest', CC iii (1911), 533.
[60] LCWTOWRC, *Statement of an amateur pitbrow worker*, Manchester n.d., 2.
[61] Ibid.
[62] WCG, 'Pit-brow lassies', CN xlii (1911), 1043.

are anxious to abolish the self-reliance, robustness and poise that comes from healthy, self-respecting labour, and that they would gradually restrict all women's work to within indoors, where they many pine, and grow rickety and anaemic.[63]

The Women's Freedom League, too, was favourably impressed by the health, vigour and joyfulness of the workers. The dignified statements made by the older women, who had inherited the work from their mothers and grandmothers, they said, 'carried much weight both as to the power of the women to do the work and the injustice of the charges against its moral effect'.[64] It was patently unfair that the fate of 5,000 women was to be decided by the votes of a few men. Even long-standing advocates of protection, such as the WIC, supported the pit-brow women's right to keep their jobs. Since, the group argued, medical evidence unanimously declared this work healthier than factory work, 'it would seem that any conditions which are objectionable could be altered by restricting men's license in foul speech, providing suitable dress for women, baths and proper sanitary accommodations, together with a better rate of pay'.[65]

Pit-brow workers appeared at many of the meetings sponsored by their supporters and played the leading role in a deputation to the Home Secretary on 3 August 1911. The forty-five women and girls from various collieries in the Wigan area who descended upon London to save their jobs were quite a spectacle as they marched along the Embankment dressed in their shawls and clogs. The mayor of Wigan, Sam Woods, his wife, and the local vicar joined numerous MPs who led them to meet Mr Masterman, under-secretary at the Home Office. Mr Harmood-Banner, MP for Liverpool, who introduced the group, claimed that this amendment stemmed not from concern over safety at the mines but from anxieties over whether or not this was a fit and proper occupation for women. The girls right before them, he said, 'showed no emaciation, no signs of anaemia, no signs of excessive work, excessive hardship, or excessive difficulty' forcing them to conclude that their work was not unsuitable for women.[66] In fact, he noted, several had left factory work in ill health which improved with their labour on the pit-brow. MP Neville waxed eloquent about their physical appearance stating that 'Those who had studied classical art would know what he meant when he said they had a perfect carriage. They walked like Juno. – (Laughter and applause). They were perfectly developed women.'[67]

Other members of the deputation defended women's work in the language of healthy motherhood; a major point of contention. Mayor Woods denied that lifting lumps of coal caused strain or serious internal injuries. Dr Cooke

---

63 NUWSS, 'The pit-brow women's protest', 533.
64 Women's Freedom League, 'The pit-brow lasses again', *Vote* v (1911), 25.
65 WIC, 'Industrial notes', *WIN* lv (1911), 137.
66 The proceedings of this meeting were reported in the MG, 4 Aug. 1911.
67 Ibid.

agreed that this work was not unsuitable for a healthy woman. 'As a class', he told Masterman,

> women employed on the pit-brow were very healthy in body, and morally they compared very well with any other class of women. . . . They were dressed in a sanitary fashion throughout and the head shawls they wore kept the dust from their hair, while the rest of their bodies were amply protected. Strain and hernia were practically unknown to them. They were robust in appearance, and those women married became the mothers of children of a far healthier type than those of other women workers.[68]

The argument that pit-brow women were robust and producing children healthier than those of women in other trades was a crucial one to make within this context.

The various protest meetings as well as the deputation of the Wigan 'pit-brow lasses' were widely reported in the new journalistic newspapers like the *Manchester Guardian*, the *Pall Mall Gazette*, the *Star* and the *Daily Chronicle*.[69] They carried editorials and letters on the subject, sent out special correspondents to investigate the conditions of work at the pit-brow and, in the end, enthusiastically endorsed the women's right to continue this work. It was abundantly clear that the pit-brow workers' physical appearance and manner of dress particularly fascinated them and their readers. The *Daily Chronicle's* report of the 'Deputation in clogs' included a picture of the pit-brow girls while the *Star's* article commented that the members of the deputation dwelt upon the good physical appearance of the workers. The deputised, the latter article pronounced, 'evidently realised that the good appearance of the girls was the best argument in favour of the worker; and they plunged into such glowing eulogies that brought blush to the cheeks of the girls'.[70] The women were also eulogised in letters to the editors such as one to the *Manchester Guardian* from F. Wynne who noted that Lancashire pit-brow workers were known by the scarlet band of flannel worn across the forehead to protect them from the coal dust. As they walked home from work in their clogs, he added 'you see but a triangle of it under the shawl which invariably covers her head and shoulders. It is a curiously becoming headdress, and one would not willingly miss from our monotonous streets the sight of these light-hearted homecoming girls. For their superior health and vigour is no fairy tale'.[71] He attributed their good health to outdoor work which did not involve 'any kind of strain, either muscular or nervous, that is particularly injurious to the physique of women'.[72] Wynne's description, in particular, evokes the sense of

---

68 Ibid.
69 See, for example, ibid. 3, 4 Aug. 1911; PMG, 4 Aug. 1911.
70 *Star*, 4 Aug. 1911.
71 This letter was quoted in CN xlii (1911), 1043.
72 Ibid.

carnival that was often conveyed in the descriptions of the spectacle of the pit-brow women 'on parade'.

The combination of women's groups, the press, medical opinion and governmental support ultimately defeated Markham's clause in December 1911. Its opponents presented clear evidence that work on the pit-brow was not injurious to women or their children. Moreover, the diverse investigations into the conditions of women's work and experiences in mining areas revealed that many men in the trade objected to women's work because they thought it unsuitable; it offended their sensibilities.[73] Under-secretary of state Masterman told the pit-brow deputation that the government opposed Markham's amendment and promised that it would attempt to secure its deletion or present an alternative clause defining working conditions on the pit brow. Thus, the resulting legislation included a vague clause which forbade lifting, carrying or moving anything so heavy as to cause injury to boys, girls or women.[74]

Since the late 1880s the splintered feminist movement had pursued different strategies to improve the industrial and economic position of working women. Social feminists called for increased state intervention for exploited women who were unable to remedy their situation by self-help via unionisation. They claimed that the opposition of equal-rights feminists to protection overlooked the historical and actual position of women in the labour market.[75] Meanwhile, groups like the SPW and the WIDC believed that the women associated with the trade union/labour movement betrayed their sex by allying with their male counterparts. Their divergent views on protection stemmed from their belief in the primacy of either class or sex solidarity. After 1903 the various suffrage groups founded by Gore-Booth and Roper presented an alternative position on the issue of protection. The working-class members of these organisations, drawn mainly from the textile mills, had benefitted from half a century of protection and had found that unionisation was not enough to improve their position *vis-à-vis* men. During a period when industrial issues had, in fact, become national political issues, they viewed the disenfranchisement of women as a serious cause around which to rally working womens' grievances. Thus, they believed that women, regardless of their views on protection, should agitate for the vote so that they could decide whether or not it was necessary and, if so, its particular form.

---

[73] This point was made, for example, by Stephen Walsh MP from Wigan at the 3 Aug. deputation. He cited the recent resolution against women's work on the pit brow passed by the Scottish Miner's Federation. The group's president, Robert Smillie, had argued that this was not suitable work for mothers of the future generation. See MG, 4 Aug. 1911.
[74] This clause was added after an extensive, and interesting, discussion in parliament: *Hansard*, 5th ser. xxxii. 1234–80 (5 Dec. 1911).
[75] This argument was made, for example, in Hutchins and Harrison, *A history of factory legislation*, 197–9.

Their differences on the general subject of protection were amplified with the emergence of the new avenue of state intervention: dangerous trades regulations. It became a particularly divisive issue because the enactment of such measures came to revolve around the protection of women in order to safeguard their ability to produce healthy children. In so doing, this legislation centred around this central aspect of sexual difference. This, in turn, forced feminists to confront the difficult issue of reconciling the dual roles of women who were workers and, in many cases, mothers. Reflecting upon this issue, MacDonald and her associates in the WLL asserted that working mothers were a 'necessary evil' since many families needed their additional income.[76] But they would always be secondary workers in the labour force. As she said in a lecture, 'With a man his outside employment is his life work; with a woman it has to share her time and energies with her home duties. Marriage must always make the economic position of women very different from that of men.'[77] From her perspective, women and men were different and unequal. Such views were in line with those expressed by many of her contemporaries such as Dr Oliver.

In stark contrast, Gore-Booth and her organisations ignored sexual difference and repeatedly made the radical suggestion that women had a right to work. From her perspective, the whole issue of dangerous trades regulations was a moot one until women were allowed to participate in its construction. For her, women faced greater danger from male trade rivals and 'shortsighted philanthropists' like MacDonald who sponsored protection. As she wrote in *Women's right to work*, 'When men say that the employment they covet for themselves in too unhealthy or unpleasant or degrading for women, behind their "chivalry" you will find masquerading the ancient primitive desire to get something for oneself, cost what it may for others.'[78] In her mind, trade rivalry was at the root of the campaigns for legislation to protect women from allegedly dangerous work. People rejoice at the idea of protection, Gore-Booth also wrote in an article entitled 'No trades for women', whenever they feel vaguely sentimental or uncomfortable about the bad working conditions for women. Thus, she said in a pointed reference to social feminists like MacDonald:

> when you vaguely think it must be horrid to be a barmaid, and spend your life in a public house, or it must be dreadful to walk on the tight-rope, or sort coal at the pit-brow, or hammer chains in a forge, you gather your friends together

---

[76] This opinion was expressed, for instance, in Mrs Player and Mrs MacDonald, 'Wage earning mothers', *League Leaflet* iii (1911).
[77] 'That the economic position of women can be best improved by legislation', typescript in the letters and papers of Mrs J. R. MacDonald, British Library of Political and Economic Sciences, London School of Economics, vol. ii. 102.
[78] Eva Gore-Booth, *Women's right to work*, Manchester [1909], 3.

and start a Society to do away with or 'abolish' barmaids, or circus-riders, or acrobats or pit-brow workers, or chain makers.[79]

Unfortunately, she argued, lack of practical knowledge about the lives of working women made such campaigners oblivious to the fact that their societies promoted legislation that would result in unemployment for thousands of women. And, Gore-Booth asserted, 'When no married women are able to earn a penny, either in a factory or in their homes, the suffering entailed on them and their children is simply unthinkable.'[80] She firmly believed that women, like men, had a right to work and to earn a living.

In a similar manner, members of FOLD or the WIDC asssserted equity and equality in the workplace. In 1903 Vynne and Blackburn of FOLD wrote that 'Every possible precaution against danger of any kind should be made irrespective of sex. Whenever precautions are obligatory in the case of women, an employer is simply encouraged to discontinue the employment of women, when such employment obliges him to keep his premises in a more healthy state.'[81] They vehemently asserted that men needed just as much protection as women because 'the health of men is as important to the state as the health of women'.[82] However, they made this argument at precisely the time when new arguments about difference had been introduced into the discourse of protection. Their arguments for equity and equality were diametrically opposed to new medical theories that emphasised women's greater risk of contracting diseases like lead poisoning. Even more important, during the period from the 1890s to the outbreak of the First World War, women's roles as mothers were accentuated as never before. And it was not social motherhood but rather biological motherhood that was being monitored and protected through dangerous trades regulations. Concerns about the future of the race also meant that, as potential mothers, their possible rights were secondary to those of the state and their unborn children.

Despite their ideological differences, however, all these groups of women attempted to influence the public dialogue on women's work, motherhood and the issue of how far the state should intervene in the working and private lives of women workers. Through their tireless efforts in gathering and disseminating information, through the printed and spoken word, they were agents in the making of protection despite their political exclusion.

---

[79] Idem, 'No trades for women', 513–14.
[80] Ibid. 514.
[81] Blackburn and Vynne, *Women under the Factory Act*, 84–5.
[82] Ibid.

Epilogue

In this epilogue I want to make some concluding remarks about the creation of dangerous trades regulations and suggest that the diverse efforts to protect unborn children in the lead trades were an important, but historically neglected, facet of a larger pre-war campaign to safeguard the 'future of the race'. During the years after 1900 various historical actors regarded motherhood as a major political issue and linked it to wide-ranging concerns about racial deterioration and its consequences for the nation and empire. In addition, I will place these developments within the international context and will briefly explore a similar recasting of the woman worker problem in pre-war France and Germany.

Between 1830 and 1914 a discourse of danger dominated the public discussion of women's work in textile factories, nail and chain making workshops, white lead and pottery firms, on the pit brow of coal mines and even behind the counter of a bar; in this public debate, different kinds of danger were emphasised at different moments. Opponents of women's work in textile factories repeatedly expressed their concern about the moral and sexual implications of such labour. Many contemporaries shared Peter Gaskell's belief that factories were 'hotbeds of lust'. Maternal misconduct replaced sexual misconduct as the charge levelled against working women during the early 1870s. Absent from home and plying their children with artificial food and cordials, working mothers were responsible, doctors argued, for the high infant mortality rates in manufacturing districts. By the 1890s the concerns had shifted, yet again, to the impact of certain kinds of work on women's reproductive organs. Arguments for the protection of women workers, through either the regulation or the prohibition of their labour, followed these diverse representations of the hazards of their work. Convinced of the detrimental impact of lead work on maternity, the government created the legal apparatus for the regulation of dangerous trades in 1891 and 1895. As a result, the Home Secretary was empowered to restrict or prohibit the work of certain persons, women and children, from certain kinds of work in dangerous trades. In 1898 women were banned from working in the white beds, rollers, washbecks, stoves or in packing white lead; they were also subjected to monthly medical examinations, with suspension from work if ill, in the pottery trade.

This complex episode in social policy-making reveals that Home Secretaries and government officials responded to external pressure from a variety of groups. Sensational *exposés* and editorials, particularly those in the *Daily Chronicle*, generated 'the steam known as public opinion' which was, according to W. T. Stead, the greatest force in modern politics.[1] By uncovering and publicising the hazards of the white lead and pottery trades, the new journalistic press initiated the public dialogue on dangerous trades. Their investigative reporting also accentuated medical opinions about lead poisoning and gave coverage to activities of the different organisations lobbying for or against regulation for women in hazardous trades. Pottery manufacturers, diverse feminist organisations and, to a lesser extent than in previous episodes, working men participated in the public dialogue. Middle- and working-class women were a visible and vocal presence as they undertook investigations of women's work, held public meetings and sent petitions and deputations to the Home Office. The WTUL, I have suggested, supplanted working men as the chief advocates of special labour laws for women during the period under consideration. They were joined by women also affiliated with the labour movement in the WLL. Because men and women were different and had different needs, Margaret MacDonald has argued, women needed special state protection. Women in the SPW, the WIDC and the FOLD vehemently disagreed with this general perspective and the specific idea that drastic measures were required to safeguard the health of women and their unborn children in dangerous trades. Their viewpoint was, of course, a minority one in a society founded on the belief in fundamental natural differences between men and women.

Ideas about femininity and masculinity led to a gendered notion of dangerous work. One of the main concerns of this book has been the emergence of the novel idea that certain types of labour were especially dangerous for women. Women were, according to Dr Thomas Oliver and other doctors, more susceptible to lead poisoning than men. Convinced that lead work interfered with the ability to bear healthy children, these doctors repeatedly urged the government to enact regulations to safeguard the potential offspring of working women. There is no doubt that this labour was hazardous to women's health and reproduction. However, the theory of women's greater susceptibility to lead poisoning is questionable, and alternative solutions to this problem of industrial illness did exist. The rigorous enforcement of existing special rules for the trades is the most obvious action that could have been taken to safeguard the health of women workers. Moreover, and this has been a major contention of this book, work in the white lead and pottery trades was also dangerous for men and affected their reproductive efforts. But contemporary ideas about masculinity and work limited state intervention in their working lives. Male potters objected to medical examination, with possible suspension from work, unless they were awarded compensation. Their argument was that this form of protection would have been a serious infringement of their ability to fulfil their natural role as the family breadwinner.

[1] Stead, 'Government by journalism', 661–2.

Home Office officials quickly dismissed the WEU's suggestion that 'the legal protection for workers be equal for both sexes'.[2] The state's firmly entrenched resistance to equal treatment for men and women also came through clearly in Home Secretary Asquith's response to the WIDC's arguments against special laws for women in 1895. He reminded them that the central premise of protection was that 'public safety and public interest' required the different treatment of women and men. And, he added, 'An argument based upon the assumption that there was no distinction between the two sexes and that they ought to be treated alike was an argument fatal, not only to the Bill now before Parliament, but to the whole series of Factory Acts from the very beginning.'[3] In the debate over dangerous trades regulations, participants continually emphasised that women bore children and this fact justified the state's intrusion into their private lives. As Principal Lady Inspector Adelaide Anderson remarked 'The special interest of the whole community in the protection of maternity and the health of young workers, for example, was the chief point on which we had influence on developing regulations for the white lead trade.'[4] In the end, the state enacted measures to protect their potential offspring even when women claimed that their work did not interfere with childbearing or when they pleaded economic necessity. These were small concerns compared to the larger, and more important, one of serving the 'public interest'.

In the years after 1900 there were other calls for the regulation of women's work to serve the interests of the race and nation. In the political world, John Burns and others argued that women's work was a threat to motherhood and placed children at risk. Burns, who led a deputation on behalf of pottery workers, expressed his opinions about married women's work on numerous public occasions. At a political meeting at Caxton Hall in 1908, for example, he summed up his perspective in a pithy phrase: 'A mother cannot sub-let her maternity.'[5] He also spoke of the dangers of married women's work at national conferences on infantile mortality held in 1906 and 1912. He told his audience at the latter conference, patronised by King George and Queen Mary, that

> the stream of life is no purer than its source. The source is motherhood. Purify, dignify, and glorify motherhood by every means within your power. Exalt the mother and you elevate the child . . . address yourself to the health of the infant, the happiness of the mother, and the invincible and undying vigour of the great race to which we belong.[6]

---

2 *Third report of the Women's Emancipation Union*, 30.
3 *Times*, 21 June 1894.
4 Anderson, *Women in the factory*, 114.
5 This speech was quoted in William Anderson, 'Wives as wage earners', *Women Worker* xi (1908), 277.
6 *Report of the proceedings of the National Conference on Infantile Mortality*, London 1912, 17.

Worried about 'the health of the infant' and the 'vigour of the great race', he suggested restricting married women's work, without pay, three months before and six months after childbirth.

Most people who spoke of the threat of women's work to the future of the race proposed, like Burns, the extension of the mandatory one month maternity leave after birth to several months prior to birth. And the common justification was that such action would allow women to complete the important act of motherhood. Very significantly, many proponents of such measures began to argue that it was not only the state's duty to protect the unborn, its future citizens, but also its right. Such intervention in the lives of working mothers would serve the 'public interest'. Dr W. M. Hamilton made such an argument in his 1908 presidential address before a branch of the Society of Medical Officers of Health. Most people, he asserted, had recognised the principle that 'it is the right and duty of the State to restrict the freedom of the individual in the interests of the community'.[7] He felt that further restrictions on the labour of pregnant workers were warranted because their desire to supplement the family income had led them to disregard the 'injury done to the unborn child'.[8] Thus he suggested a ban of four to six months before birth, again, with no mention of monetary compensation, in order reduce the infant mortality problem, 'one of the greatest blots on our twentieth century civilization'.[9] 'You may expel nature with a pitchfork', he concluded, 'but she will return in the dwarfed, rickety, undersized, and physical degenerate of the second and third generation.'[10] This justification for state intervention was also supported by May Tennant, a former factory official, in 1902. 'That there is danger to the race', she wrote, 'in the engagement in factory life of the mothers of young children, should be beyond challenge; always danger to the child, often danger to the mother; and sacrifice of infant life, failure of infant promise follow, have followed, and must follow, as surely as the leaves follow the frost.'[11] She also contended that 'The freedom to labour is no sacred right when its exercise involves injury to others.'[12] Dr Percy Frankland, who had supported Oliver's proposed ban on women who worked with lead in the pottery trade, agreed with this line of argument. As he said to his fellow members of the Sanitary Institute in 1899: 'We, as members of this Congress which is devoted to the promotion of public health, must be agreed that any rights, either of men or women, which are opposed to the welfare of the future generations, must be sacrificed without hesitation or remorse.'[13] While

[7] Dr W. M. Hamilton, 'Some points in industrial hygiene', *PH* xxii (Oct. 1908–1909 Sept.), 125.
[8] Ibid. 127.
[9] Ibid.
[10] Ibid.
[11] May Tennant, 'Married women's work and infant mortality', in Oliver, *Dangerous trades*, 73–84 at p. 73.
[12] Ibid. 83.
[13] Frankland, 'Address to section iii', 389.

including both sexes in this public health manifesto, he only discussed the deprivation of women's rights in order to protect future generations.

These representations of the woman worker, with the emphasis on protection of motherhood in the interest of the nation, were part of the larger pre-war preoccupation with population. As England became increasingly caught up in economic and imperial rivalry, the quality and quantity of its population became an urgent political issue. And, at the same time it appeared that its population was declining on both counts. The Boer War revealed to a stunned nation the poor physical condition of volunteers from the working class, one-third of whom were rejected as physically unfit to fight.[14] Moreover, the nation's poor military performance against the Boers as well as its relative industrial decline *vis-à-vis* Germany made many envisage military as well as economic defeat at the hands of its chief rival. The solution was to improve the physical and industrial efficiency of the working class by focusing on its women and children. Government statistics added further alarming information. The registrar-general's report for 1907 emphasised that while the death rate for children aged one to five had fallen by 33 per cent in the previous forty years, the rate for infants under one year remained as high for the decade of the 1890s as it had been in the 1860s. This, coupled with a decline in the birth rate, meant that by 1900 fewer babies were being born and a high proportion of them were dying in their first year.[15]

Anna Davin and others have explored the implications of this preoccupation with population for women of the period.[16] As Davin has argued in her pathbreaking article on motherhood and imperialism, 'Population was power. Children, it was said, belonged "not merely to the parents but to the community as a whole"; they were "a national asset", "the capital of the country"; on them depended "the future of the country and the Empire"; they were "the citizens of tomorrow".'[17] By the early twentieth century, a powerful new idea of motherhood, she added, had begun to emerge in which it was the 'duty and destiny of women to be the "mothers of the race" '.[18] This ideology, Jane Lewis has further written, 'persuaded married women that their role in the home was of national importance and that motherhood was their primary duty'.[19] The need for healthy mothers of robust children led to what Ellen Ross has called the discovery of mothers by social thinkers.[20] Concerned

---

[14] For more on the impact of the Boer War and the movement for national efficiency see G. R. Searle, *The quest for national efficiency: a study in British politics and political thought, 1899–1914*, Berkeley 1971, 34–106.
[15] These statistics are from Dyhouse, 'Working class mothers', 73.
[16] See Davin, 'Imperialism and motherhood'; Lewis, *The politics of motherhood*, and 'The working-class wife and mother'; Dwork, *War is good for babies*; Jenson, 'Gender and reproduction'; and Michel and Koven, 'Womanly duties'.
[17] Davin, 'Imperialism and motherhood', 10.
[18] Ibid. 13.
[19] Lewis, *The politics of motherhood*, 244.
[20] See Ross, *Love and toil*, 3–10, 195–221. She does note carefully that previously there had

medical men, civil servants and politicians ascribed the problem of infant mortality to mothers who were ignorant about childcare. There was, as she, Davin and Lewis have shown, a proliferation of maternal and child welfare schemes, such as schools for mothers or crèches, aimed at eradicating that harmful ignorance.

In the three decades prior to the First World War, there was a plethora of literature produced on the problem of the 'woman worker'. Newspaper stories, investigations by factory inspectors, special governmental committees, medical men and numerous feminist organisations disseminated shocking information about the dangers of women's work outside the home. The subject matter was not new, of course, but this literature was distinguished from earlier nineteenth-century writings by its sheer volume, focus and tone. The reader easily detects the tone of urgency as authors vociferously propose their solution to this grave problem afflicting the body politic. The female body and the body politic were fused in many minds; the degeneracy of the former was symptomatic of the degeneracy of the latter. And, as this book has shown, the association of dangerous work with women's impaired procreative abilities was not limited to trades officially designated as dangerous, but extended to all trades.

Fin-de-siècle France and Germany

These kinds of concerns about race, motherhood and women's work had an international dimension as well. This was reflected, most clearly, in the passage of numerous sex-specific labour laws in both France and Germany. As was the case in England, the new protective measures encompassed not only the regulation of hours of work but women's work in dangerous trades and before and after childbirth. The working day for French women was limited to eleven hours in 1892; it was reduced to ten hours eight years later. In 1892 French legislation banned women from night work and prohibited women from employment in unhealthy or dangerous workplaces. By 1900, according to Mary Lynn Stewart, they were banned from sixty-two trades.[21] About 75 per cent of those trades produced volatile substances or used dangerous substances such as lead or mercury. Maternity legislation passed in 1913 allowed pregnant women the right to take a break from paid employment without penalty. A four-week break before birth was optional while an obligatory four-week break after childbirth was mandatory. The state provided

---

been organisations and activities in London concerned with the conditions of motherhood among the poor but clearly differentiates between them and developments during the early twentieth century. She argues that the difference is most clearly seen in the exclusive focus on mothers and child care, the sheer size and intensity of organisations, and, most important, the coercive laws which accompanied this activity.

[21] Stewart, *Women, work, and the French state*, 162.

monetary allowances to help women during their absence from work. German legislation from 1878 prohibited women from working in mines and provided for a three-week maternity leave after childbirth. A revision of that labour code in 1891 excluded women from night work, reduced their working day to eleven hours and expanded the maternity leave from three to six weeks with a doctors' note enabling a return to work after only four weeks. Lastly, the 1908 labour code mandated a ten-hour working day on weekdays with an eight-hour limit on Saturdays and barred pregnant women and new mothers from working two weeks before and six weeks after childbirth. The new mother, it should be added, had to submit a doctor's report assessing her postpartum recovery and bodily fitness before she could resume her employment.[22]

Elinor A. Accampo has examined the context within which the French legislature enacted its measures restricting women's work. She contends that three factors in late nineteenth-century France combined to influence significantly the way that men with political power viewed women: industrialisation, the decline in the rate of population growth and the culture of republicanism. The numbers of French women in industry increased steadily from about 24 per cent in 1850 to about 43 per cent in 1920. And, just as these figures were rising, so too were the rates of infant mortality in industrial cities. For instance, the rates for the *département* of the Nord, the centre of textile production, reached their peak between 1893 and 1900. This problem, moreover, was part of a larger demographic crisis which has been labelled 'the crisis of depopulation'.[23] Between 1888 and 1901 France was experiencing a minimal population growth of about 3 per cent while Germany's population, for instance, increased by about 20 per cent. This development evoked considerable fears and anxieties about the degeneration of France. The spectre of its humiliating defeat in the Franco-Prussian war loomed large in many minds and there was a sense that, in the event of a future military contest, a weakened France would not be able to contend with its enemy Germany. There was also, Accampo has shown, a critical ideological development within republican culture: the solidarism of Léon Bourgeois. Solidarism, she has written,

> posited that individuals and social classes were mutually dependent and had social obligations that rose above individual interests. Solidarism provided a rationale for restricting the *laissez-faire* economics in order to establish social reform, ameliorate class relations, and prevent social conflict.[24]

[22] Canning, *Languages of labor and gender*, 211.
[23] The tremendous anxiety about the declining birth rate in France has been amply illustrated in Karen Offen, 'Depopulation, nationalism, and feminism in *fin-de-siècle* France', *American Historical Review* lxxxix (1984), 648–76.
[24] Elinor A. Accampo, 'Gender, social policy, and the formation of the Third Republic: an introduction', in Accampo, Fuchs and Stewart, *Gender and the politics of social reform*, 1–27 at p. 14.

In particular solidarists believed that working-class women, at home and versed in household management, could stabilise the social order.

Because of these developments, she has argued, women's reproductive functions assumed a new social importance. Concern for their procreative abilities became one of the most important factors that led to the enactment of numerous measures restricting women's work. Accampo is one of several French historians who have asserted the centrality of women's reproductive roles to Third Republic politicians and their legislative initiatives. They have illustrated the pervasive discourse on the maternal body in the political debates, the motives and activities of key advocates of protective legislation, particularly doctors, and the relationship of this legislation to ideas of *laissez-faire* and a gendered conception of citizenship. They have also shown the connection between women's work and the maternal body and nationalism, national defence and the future of the race. I will briefly highlight some of their major points.

Mary Lynn Stewart and Rachel G. Fuchs, for instance, have demonstrated the decisive influence of doctors upon the creation of sex-specific legislation. Stewart has analysed the protracted debate over the provision of a maternity benefit that began in the 1880s and was not resolved until 1913.[25] During the ten-year senate debate prior to the 1913 legislation, she has noted, politicians drew upon the medical literature on married women's work and infant mortality which had been rapidly expanding during the 1890s.[26] Paul Strauss, a solidarist and the chief advocate of this legislation, employed the new medical and demographic data to convince his associates that maternity leave would be beneficial to the state. The scientific and statistical evidence used to buttress his position went virtually unchallenged. This is not surprising, however, since doctors had become the experts on the subject of infant mortality. Assessing their overall influence on sex-specific legislation, Fuchs has argued that their political power in France was stronger than in any other European country. They used, she has maintained, the depopulation scare to further their influence and argued that the regulation of motherhood and the health of the future generation was a top political priority. Consequently, to quote Fuchs, medical works became 'new weapons in the arsenal of political intervention in motherhood and childbearing'.[27] Doctors had a direct impact on legislation through their presence in the French legislature: between 1870 and 1914 they constituted between 10 and 12 per cent of its membership.[28] They were thus disproportionately represented in the legislative body and the parliamentary committees that delved into critical issues of health, reproduction and possible state social policies.

---

[25] Stewart, *Women, work, and the French state*, 169–90.
[26] Ibid. 178–85.
[27] Rachel G. Fuchs, 'The right to life: Paul Strauss and the politics of motherhood', in Accampo, Fuchs and Stewart, *Gender and the politics of social reform*, 82–105 at p. 84.
[28] Accampo, 'Gender, social policy, and the formation of the Third Republic', 15.

Strauss solidified his case for a maternity leave through the deployment of medical literature and an appeal to national interest. His nationalist rhetoric and patriotic appeal struck a chord with many of his political contemporaries preoccupied with the depopulation crisis.[29] He called upon patriots and philanthropists to cooperate in what he termed the 'battle for national defense'.[30] Since babies were needed for the country's defence and maintenance, he would tell them, maternity leave would save them for 'the national advantage'.[31] He and fellow deputy Fernand Engerand frequently drew parallels between men being called up for military service and women being called to render their maternal service. Very significantly, Strauss's appeal signified an important shift in focus: the state, rather than mothers and their babies, became the chief beneficiaries of maternity leave.

In the end, references to safeguarding the future of the race resulted in a political consensus about intervention in the lives of working women. Because of tremendous anxiety about the effect of women's labour on the birth of children, Judith F. Stone has written, 'Politicians of both the right and left called on the state to act in defense of this most fundamental "natural resource".'[32] Even radical politicians who were solidly committed to individual rights supported intervention in the lives of working women. This is clearly reflected in remarks, such as those by radicals Paul Pic and Leon Bourgeois. The former argued that

> If women who worked to excess would injure only themselves, it might be permissible to argue that the legislator should not intervene; but this is not the case. The woman injures the child she might produce. Without regulation of female labour, society will be menaced by bastardisation of the race.[33]

Bourgeois, prominent among those who identified the declining birth rate as one of France's most grave and menacing problems, defended labour legislation because it had 'the higher goal to ensure the vigour and the future of the race, by organizing social hygiene'.[34] Stone has shown that they and other radicals buttressed their position by arguing that the fact that women were mothers or potential mothers meant that they need not be considered individuals. This meant that the state could legitimately regulate their labour because it could jeopardise the lives of the children they might produce and, ultimately, the future of the French nation.

---

[29] Stewart has argued that this linkage of depopulation to nationalist issues was common in the pre-war years: *Women, work, and the French state*, 173.
[30] Fuchs, 'The right to life', 88.
[31] Stewart, *Women, work, and the French state*, 185.
[32] Judith F. Stone, 'Republican ideology, gender, and class: France, 1860s–1914', in Frader and Rose, *Gender and class*, 238–59 at p. 249.
[33] Ibid. 250.
[34] Ibid. 251–2.

Paul Weindling has demonstrated the widespread perception of a demographic crisis in Germany in the decades before the First World War. For, although the national population continued to rise, the birth rate fell by 70 per cent between 1871 and the 1930s.[35] What heightened public concern was a high infant mortality rate; at the end of the nineteenth century 20 per cent of the babies born to working-class families died in their first year.[36] There followed, he has written, 'an emotive campaign to raise the birth rate with the idealization of motherhood, of large families as "child rich", and of nationalistic appeals that Germany was a "nation without youth" '.[37] Moreover, a distinctive vision of the German mother was taking shape. The Prussian medical official Eduard Dietrich, for example, insisted that a German mother preferred breast to bottle feeding her infant and was prepared to devote herself to her family and her nation. He contrasted this self-sacrificing and caring mother to the egotistical French mother.[38] This concern for children intensified, Weindling has argued, with the great power rivalry and the need to replenish the industrial labour force. Support for the resulting infant welfare campaign moved well beyond medical and philanthropic associations to become a nationalist and imperialist movement by 1914. It encompassed organisations and politicians spanning the political spectrum as well as the government and members of the royal family. And, whatever the causes of infant mortality there was, he has concluded, the tendency to blame mothers, particularly working mothers, for this problem plaguing the nation.[39]

Kathleen Canning's work has examined more fully the connection between women's employment in the industrial labour force and the population problem in pre-war Germany. She has analysed women's participation in the workforce, the discourse on the female worker, and legislative initiatives to protect women, particularly pregnant women and new mothers. This period was characterised by the rapid industrialisation of Germany and a steady and perceptible influx of women into the workplace. What most alarmed observers was the fact that the number of married women working outside the home doubled between 1892 and 1907; as a percentage of all women workers this meant a rise from 21 to 29 per cent between 1895 and 1899 alone. This trend led, she has written, to 'the spectre of "displacement" of male workers by women or of lowered wages and "feminization" of the factories'.[40] These developments, along with anxieties about the growth of Social Democracy, industrial unrest and imperial expansion during the 1880s

---

[35] Paul Weindling, *Health, race, and German politics between national unification and Nazism, 1870–1945*, Cambridge 1989, 189.
[36] This statistic is from Canning, *Languages of labor and gender*, 196.
[37] Weindling, *Health, race, and German politics*, 189.
[38] Ibid. 204.
[39] Ibid. 196.
[40] These statistics and the analysis have been drawn from Kathleen Canning, 'Social policy, body politics: recasting the social question in Germany, 1875–1900', in Frader and Rose, *Gender and class*, 211–37 at pp. 225, 220.

and 1890s, combined to create the new social question: the problematic woman worker. According to anxious social observers, women's work endangered themselves and their families. This fear was the catalyst to the production and dissemination of a wide variety of narratives about female factory work. These narratives of danger included academic and scientific texts, state investigations, factory inspector reports as well as union, feminist and employer tracts.

A groundswell of social pressure led the government to consider and enact special laws to resolve this pressing social problem. In the end, it enacted the 1891 and 1908 labour codes and instigated an official inquiry into married women's factory employment in 1898–9. The latter was stimulated by a vigourous campaign for a legal ban on such employment. This was, Canning has argued, linked to a new discourse on women's work which appeared after the publication of a study by Rudolf Martin, a minor barrister from Saxony, in 1896. He investigated factory and married life among female textile workers in that area and concluded that women's labour would ultimately 'ravage the social body'.[41] He particularly drew attention to the impact of such employment on infants, claiming that 'it spoils the human material and damages the labour power of the nation'.[42] Canning has argued that this work marked a turning-point in the discussion of women's work; it represented a shift in emphasis from protective measures for women to a consideration of a legal ban on married women's work.

Two years later the government sponsored an official investigation of the subject. Factory inspectors devoted two years to the collection of information about the number of married women working outside the home, why they did so, and the possible consequences of a ban on their labour. While the inspectors concluded that such a course of action would be beneficial to the women and their families, they argued that it would not be a practical course of action. The fact that their families relied upon their income was noted but employers' objections seemed to carry greater weight. They had stressed that it would be difficult to replace the women, especially since they were more suited to some jobs than men. The inspectors recommended the creation of state schools where factory women could be taught sewing, cooking and infant care.

Canning aptly argues that even though the ban never became law its discussion had important consequences for women. Her reading of the factory inspector reports suggests the immense implications of the discursive exercise. Through their commentary and suggestions, Canning has concluded, factory inspectors reinforced women's secondary role in the labour market and the desire to implant within them proper domestic skills. Moreover, they compiled new and compelling evidence about the dangers of factory work to

[41] Ibid. 228.
[42] Ibid.

pregnant women. Medical men, health insurance officials and employers reported to them extraordinarily high rates of miscarriages, premature births and infant mortality. 'In specifying the site of intervention as the maternal body', she has argued, 'they began to widen the scope of the social question from the factory to the nation, to link more explicitly the conditions of work to the conditions of [national] reproduction and the national birth rate.'[43]

In England, France and Germany the powerful idea that women's work was dangerous for the state was articulated in public and parliamentary discourse. Most significantly, this idea was translated into legislation primarily passed, as its proponents made abundantly clear, to serve the political interests of the state by protecting future citizens. The fears of the English Vigilance Association, expressed during the 1870s, were confirmed as states acted upon 'the maxim that a woman is merely a piece of child-bearing mechanism, and that all faith in her affection and all regard for her rights are to be set aside out of consideration of her offspring'.[44] This maxim had important consequences for women both then and now.

---

[43] Ibid. 233.
[44] *Manchester Examiner and Times*, 20 June 1874.

Bibliography

Unpublished primary sources

London, British Library
Francis Place newspaper collection

London, British Library of Political and Economic Sciences, London School of Economics
Letters and papers of Mrs J. R. MacDonald

London, Labour Party Archives, Transport House
Women's Labour League conference reports, 1906–14
League Leaflet

London, Public Record Office, Kew
HO 45/9308/12500 Hours of women and young people in textile factories, 1872–8
HO 45/9773/B1508 Factory and Workshop Act (1878) amendment, 1887–94
HO 45/9794/B5090E Factory and workshop bill, 1891
HO 45/9848/B12393A White lead poisoning, 1892–8
HO 45/9851/B12393E Potteries, lead poisoning, 1892–1900
HO 45/9856/B12393AC Dangerous trades, white lead regulations, 1896
HO 45/9881/B16265 Factory and workshop bill, 1894
HO 45/9889/9890/B17300 Factory and Workshop Act (1891), amendment bill, 1895
HO 45/9933/B26610 Potteries, appointment of lady inspector, 1898
HO 45/10117–10121/B12393P Dangerous trades, pottery and china, 1898–1904
HO 45/10312/124109 The employment of barmaids, 1904–8
POWE 8 H/S 164/206 Coal mines, 1911

Published primary sources

Annual reports of the Trades Union Congress, Manchester 1887–99; London 1900–14

Official documents and publications
Report of the medical officer of the privy council for 1861, PP 1862, xxii
Report to the Local Government Board on proposed changes in the hours and ages of employment in textile factories, PP 1873, lv

Thirty fourth annual report of the registrar-general of births, deaths, and marriages in England, PP 1873, xx

House of Lords select committee on the sweating system: fifth report; with the proceedings of the committee, minutes of evidence, and appendix, PP 1890, xvii

Royal Commission on Labour: the employment of women, reports by Miss Eliza Orme, Miss Clara E. Collet, Miss May Abraham and Miss Margaret Irwin, lady assistant commissioners, on the conditions of work in various industries in England, Wales, Scotland, and Ireland, PP 1893–4, xxxvii

Report of the departmental committee appointed to inquire into the conditions of labour in the various lead industries, into the dangers to the workpeople employed therein, and to propose remedies; with evidence, appendices, and index, PP 1893–4, xvii

Report of the departmental committee on the conditions of labour in the Potteries, the injurious effects upon the health of the workers, and proposed remedies, PP 1893–4, xvii

Royal Commission on Labour: third report, digest of evidence, group c, PP 1893–4, xxxiv

Report of the chief inspector of factories and workshops for 1897, PP 1898, xiv

Report on the employment of compounds of lead in the manufacture of pottery, their influence upon the health of the workpeople, with suggestions as to the means which might be adopted to counter their evil influence, PP 1899, xii

Report of the chief inspector of factories and workshops for 1906, PP 1907, x

Order from Home Secretary Herbert Gladstone, 22 May 1907, PP 1907, xviii

Report of the departmental committee appointed to inquire into the dangers attendant on the use of lead: the danger or injury to health arising from dust and other causes in the manufacture of earthenware and china, and in the processes incidental therein, including the making of lithographic transfers, PP 1910, xxix

Debates of the House of Commons, 3rd ser. cccliv, 1891; 4th ser. viii, 1893; xxvii, 1894; xxxi, xxxii, 1895; 5th ser, xxxii, 1911

Newspapers and periodicals

British Medical Journal
Common Cause
Contemporary Review
Co-operative News
County Advertiser for Staffordshire and Worcestershire
Daily Chronicle
Daily News
Daily Telegraph
Echo
Englishwoman
English Woman's Journal
Englishwoman's Review
Evening News
Fortnightly Review
Journal of Hygiene

Journal of the Sanitary Institute
The Lancet
Link
Manchester Examiner and Times
Manchester Guardian
New Review
Nineteenth Century
Northern Star
Pall Mall Gazette
Penny Illustrated Press
Pottery Gazette
Public Health
St James's Gazette
Scotsman
Staffordshire Sentinel
Star
Textile Mercury
The Times
Westminster Gazette
Woman Worker
Women's Industrial News
Vote

Contemporary books and articles

Acton, Dr William, *The functions and disorder of reproductive organs, in childhood, adult age, and advanced life, considered in their physiological, social, and moral relations*, in Janet Horowitz Murray (ed.), *Strong minded women and other lost voices from nineteenth century England*, New York 1982, 127–9

Alderson, Dr James, 'On the effects of lead upon the system', *Lancet* mdccclii (1852), 75, 165–7, 212–14, 391–3, 416–19

Anderson, Dame Adelaide, *Women in the factory: an administrative adventure, 1893–1921*, London 1922

Anderson, William, 'Wives as wage earners', *Women Worker* xi (1908), 277–8

Arlidge, Dr John, *The hygiene, diseases, and mortality of occupation*, London 1892

――― *The pottery manufacture in its sanitary aspects*, Hanley 1892

――― 'The position of the study of industrial diseases: its past neglect and its scope', *JSI* xv (1895), 517–20

Ballantyne, Dr J. W., 'Ante-natal causes of infant mortality, including parental alcholism', in *National Conference on Infantile Mortality* (1906), 124–43

――― *Expectant motherhood*, London 1914

Blackburn, Helen and Nora Vynne, *Women under the Factory Act*, London 1903

Boucherett, Jessie, 'Events of the quarter – Mundella's bill and shop hours regulation bill', *ER* iv (1873), 209–11

――― 'Legislative restrictions on women's labour', *ER* iv (1873), 252–8

――― 'The new factory legislation', *ER* xxii (1891), 73–9

―――― 'Lead works and some other unhealthy industries', *ER* xxv (1894), 10–15

―――― and Helen Blackburn (eds), *The condition of working women under the Factory Acts*, London 1896

Collis, Dr Edgar L. and Major Greenwood, *The health of the industrial worker*, London 1921

Cooke-Taylor, Whately, 'What influence has the employment of mothers in manufactures on infant mortality; and ought any, and what, restrictions to be placed on such employment?', in Charles Wager Ryallis (ed.), *Transactions of the National Association for the Promotion of Social Sciences*, London 1874, 569–85

Dearden, Dr W. F., 'The relation of public health to industrial diseases', *PH* xxiv (1911), 209–10

Dilke, Lady Emilia, 'Trade unionism among women', *Fortnightly Review* lv (1891), 741–6

―――― *The industrial position of women*, London 1895

Frankland, Dr Percy, 'Address to section iii', *JSI* xx (1900), 386–93

Freedom of Labour Defence Association, 'Notes and incidents', *ER* xxxi (1900), 33–5

―――― 'Notes of the quarter', *ER* xxxi (1900), 94–7

―――― *Statute mongery: their results, the remedy*, London 1901

―――― 'Barmaids', *ER* xxxiv (1904), 23

Gaskell, Peter, *The manufacturing population of England*, London 1833; repr. New York 1972

Gore-Booth, Eva, *Women workers and parliamentary suffrage*, Manchester [1904]

―――― *The Barmaids' Political Defence League*, Manchester [1907/8]

―――― 'The women's suffrage movement amongst trade unionists', in Brougham Villiers (ed.), *The cause for women's suffrage*, London 1907, 50–62

―――― 'The Home Office and the florists', *CC* i (1909), 501–2

―――― 'No trades for women', *Englishwoman* i (1909), 507–17

―――― *Women's right to work*, Manchester [1909]

Greenhow, Dr Edward, *Papers relating to the sanitary state of the people of England*, London 1858

Greenwood, Florence, *Is the high infantile and death rate due to the occupation of women?*, London 1901

Hamilton, Dr Alice, *Industrial poisons in the United States*, New York 1925

Hamilton, Dr W. M., 'Some points in industrial hygiene', *PH* xxii (Oct. 1908–Sept. 1909), 125–9

Heather Bigg, Ada, 'Female labour in the nail trade', *Fortnightly Review* xxxix (1886), 827–32

Hope, Dr E. W., Dr W. Hanna and Dr C. O. Stallybrass, *Industrial hygiene and medicine*, London 1923

Hutchins, B. L. and Amy Harrison, *A history of factory legislation*, London 1903; repr. London 1970

Joint Committee on the Employment of Barmaids, *The barmaid problem*, London 1903

—— *Women as barmaids*, London 1905
Kay, Dr James, *The moral and physical conditions of the working classes in the cotton manufacture in Manchester*, London 1832; repr. London 1970
Kinnear, John Boyd, *The right of women to labour*, London 1873
Lancashire and Cheshire Women Textile and Other Workers' Representation Committee, *Statement of an amateur pitbrow worker*, Manchester n.d.
Legge, Dr Thomas, 'Industrial lead poisoning', *Journal of Hygiene* i (1901), 96–110
—— *Industrial maladies*, London 1934
—— and Dr Kenneth Goadby, *Lead poisoning and lead absorption: the symptoms, pathology, and prevention, with special reference to their industrial origin and account of the principal processes involving risk*, London 1912
MacDonald, Margaret, *Labour laws for women: their reason and results*, London 1900
Mallet, Mrs Charles, *Dangerous trades for women*, London 1893
March-Phillipps, Evelyn, 'The new factory bill: as it affects women', *Fortnightly Review* lv (1894), 738–48
—— 'Factory legislation for women', *Fortnightly Review* lvii (1895), 732–44
Massingham, Henry, *The London daily press*, London 1892
Nash, Rosalind, *Life and death in the Potteries*, Manchester 1898
National Union of Women's Suffrage Societies, 'The pit-brow women's protest', CC iii (1911), 533
Newman, Dr George, *Infant mortality: a social problem*, London 1906
—— *The health of the state*, London 1913
Newsolme, Dr Arthur, *Report on infant and child mortality*, London 1910
O'Connor, T. P., 'The new journalism', *New Review* (1889), 423–34
Ogle Moore, Helen and Edith Hare, 'Report to the Society for Promoting the Employment of Women on the work of women in the white lead trade, at Newcastle-upon-Tyne', in Boucherett and Blackburn, *The condition of working women*, 77–84
Oliver, Dr Thomas, 'Goulstonian lectures on lead poisoning in its acute and chronic manifestations', *BMJ*, 8 Mar. 1891, 505–8; 14 Mar. 1891, 571–3; 21 Mar. 1891, 627–34
—— 'Lead and its compounds', in Oliver, *Dangerous trades*, 282–372
—— 'Address to the conference on industrial hygiene', *JSI* xxiv (1904), 171–91
—— *Diseases of occupation from the legislative, social, and medical point of view*, London 1908
—— 'Lead poisoning and the race', *BMJ*, 13 May 1911, 1096–8
—— *Lead poisoning from the industrial, medical, and social point of view*, London 1914
—— *Occupations from the social and hygenic point of view*, Cambridge 1916
—— (ed.), *Dangerous trades: the historical, social, and legal aspects of industrial occupations as affecting health, by a number of experts*, London 1902

Patten, Dr Cooper, 'The ante-natal causes of infant mortality', *PH* xxiii (1910), 330–8

Player, Mrs and Mrs MacDonald, 'Wage earning mothers', *League Leaflet* iii (1911)

Prendergast, Dr W. Dowling, *The potter and lead poisoning*, London 1898

Reid, Dr George, 'Infant mortality and female labour in relation to factory legislation', *JSI* xv (1895), 497–511

—— 'Infant mortality and factory labour', in Oliver, *Dangerous trades*, 84–9

—— 'Infant mortality and the employment of married women in factory labour before and after confinement', in *National Conference on Infantile Mortality* (1906), 223–36

Report of the proceedings of the National Conference on Infantile Mortality, London 1906

Report of the proceedings of the National Conference on Infantile Mortality, London 1912

Simon, Dr John, *Public health reports*, London 1887

Society for Promoting the Employment of Women, 'Statement of the Association for the Employment of Women', *English Woman's Journal* iv (1859), 54–9

—— 'Thirty-third annual report of the Society for Promoting the Employment of Women', *ER* xxiii (1892), 182–3

—— 'Memorial to Home Secretary Matthew Ridley', *ER* xxx (1899), 204–5

Stead, William T., 'Government by journalism', *Contemporary Review* xlix (1886), 653–74

Strachey, Ray, *The cause: a short history of the women's movement in Great Britain*, London 1928

Tennant, May, 'Married women's work and infant mortality', in Oliver, *Dangerous trades*, 73–84

Thackrah, Charles Turner, *The effects of the principal arts, trades, and professions, and of civic states and habits of living, on health and longevity: with suggestions for the removal of many of the agents which produce disease, and shorten the duration of life*, London 1832 edn; repr. London 1957

The trials of Feargus O'Connor and fifty-eight others charged with sedition, Manchester 1843; repr. New York 1970

Tuckwell, Gertrude, *The state and its children*, London 1894

—— *Women's work and factory legislation: the amending act of 1895*, London 1895

—— 'Commercial manslaughter', *Nineteenth Century* (Aug. 1898), 253–8

—— *Constance Smith: a short memoir*, London 1931

Vigilance Association for the Defence of Personal Rights, *Constitution and rules*, Manchester 1871

—— *The factory (health of women) bill*, London 1874

—— *Proposed legislative restrictions upon the labour of women*, London 1874

—— *The right of women to labour*, London 1874

Women's Cooperative Guild, 'Pit-brow lassies', *CN* xlii (1911), 1043

Women' Emancipation Union, *The Women' Emancipation Union: its origin and its work*, Manchester 1892
Women's Freedom League, 'The pit-brow lasses again', *Vote* v (1911), 25
Women's Industrial Council, 'The employment of barmaids', *WIN* xxxiv (1906), 542
——— 'Industrial notes', *WIN* lv (1911), 137
Women's Labour League, *Second annual conference report* (1907)
——— *Fifth annual conference report* (1910)
Wood, George, 'The course of women's wages during the nineteenth century', in Hutchins and Harrison, *A history of factory legislation*, appendix a, 257–308

Secondary sources

Accampo, Elinor A., 'Gender, social policy, and the formation of the Third Republic: an introduction', in Accampo, Fuchs and Stewart, *Gender and the politics of social reform*, 1–27
——— Rachel G. Fuchs and Mary Lynn Stewart (eds), *Gender and the politics of social reform in France, 1870–1914*, Baltimore 1995
Alexander, Sally, 'Women, class, and sexual difference in the 1830s and 1840s: some reflections on the writing of feminist history', *History Workshop* xvii (1984), 125–49
Baron, Ava, 'Gender and labor history: learning from the past, looking to the future', in Ava Baron (ed.), *Work engendered: toward a new history of American labor*, Ithaca 1991, 1–46
Bartrip, Peter, 'The rise and decline of workmen's compensation', in Weindling, *Occupational health*, 157–79
——— 'Expertise and dangerous trades, 1875–1900', in Roy MacLeod (ed.), *Government and expertise: specialists, administrators and professionals, 1860–1919*, Cambridge 1988
——— ' "Petticoat pestering": the Women's Trade Union League and lead poisoning in the Staffordshire Potteries, 1890–1914', *Historical Studies in Industrial Relations* ii (1996), 3–26
——— and S. B. Burman, *The wounded soldiers of industry: industrial compensation policies, 1833–1897*, Oxford 1983
Blackburn, Sheila, 'Working-class attitudes to social reform: Black Country chainmakers and anti-sweating legislation', 1880–1930', *International Review of Social History* xxxiii (1988), 42–69
Boston, Ray, 'W. T. Stead and democracy by journalism', in Weiner, *Papers for the millions*, 91–106.
Boston, Sarah, *Women workers and the trade union movement*, London 1980
Brand, Jeanne L., *Doctors and the state: the British medical profession and government action in public health, 1870–1912*, Baltimore 1965
Brown, Lucy, *Victorian news and newspapers*, Oxford 1985

Canning, Kathleen, 'Feminist history after the linguistic turn: historicizing discourse and experience', *Signs* xix (1994), 368–404

—— *Languages of labor and gender: female factory work in Germany, 1850–1914*, Ithaca 1996

—— 'Social policy, body politics: recasting the social question in Germany, 1875–1900', in Frader and Rose, *Gender and class*, 211–37

Clark, Anna, 'The rhetoric of Chartist domesticity: gender, language, and class in the 1830s and 1840s', *Journal of British Studies* xxxi (1991), 62–88

—— *The struggle for the breeches: gender and the making of the British working class*, Berkeley 1995

Clark, Claudia, *Radium girls: women and industrial health reform, 1910–1935*, Chapel Hill 1997

Collette, C., *For Labour and for women: the Women's Labour League, 1906–1916*, Manchester 1989

Conway, Jill, 'Stereotypes of femininity in a theory of sexual evolution', in Martha Vicinus (ed.), *Suffer and be still*, Bloomington 1973, 140–54

Copelman, Dina, 'The gendered metropolis: *fin-de-siècle* London', *Radical History Review* lx (1996), 38–56

Cott, Nancy, 'Passionlessness: an interpretation of Victorian sexual ideology, 1790–1850', *Signs* iv (1978), 219–36

Creighton, Colin, 'The rise of the male breadwinner family: a reappraisal', *Society for the Comparative Study of Society and History* xxxviii (1996), 310–37

Cross, Gary (ed.), *A quest for time: the reduction of work in Britain and France, 1840–1940*, Berkeley 1989

Daffin, Jean and David Thoms, *Caring and sharing: the centenary history of the Co-operative Women's Guild*, Manchester 1983

Daniels, Cynthia R., *At women's expense: state power and the politics of fetal rights*, Cambridge, Mass. 1993

Davidoff, Lenore and Catherine Hall, *Family fortunes: men and women of the English middle class, 1780–1850*, London 1987

Davin, Anna, 'Imperialism and motherhood', *History Workshop* v (1978), 6–66

Dreher, Nan H., 'The virtuous and the verminous: turn-of-the-century moral panics in London's public parks', *Albion* xxix (1997), 246–67

Duden, Barbara, *Disembodying women: perspectives on pregnancy and the unborn*, Cambridge, Mass. 1993

Dupree, Marguerite W., 'The community perspective in family history: the Potteries during the nineteenth century', in A. L. Beier, David Cannadine and James M. Rosenheim (eds), *The first modern society: essays in English history in honor of Lawrence Stone*, Cambridge 1989, 549–73

—— *Family structure in the Staffordshire Potteries, 1840–1880*, Oxford 1995

Dwork, Deborah, *War is good for babies and other young children: a history of the infant and child welfare movement in England, 1898–1918*, London 1987

Dyhouse, Carol, 'Working-class mothers and infant mortality in England, 1895–1914', *Journal of Social History* xii (1978), 73–98

Easlea, Brian, *Science and sexual oppression*, London 1981

Fee, Elizabeth, 'Science and the woman problem in historical perspective', in M. S. Teitelbaum (ed.), *Sex differences: social and biological perspectives*, Garden City, New York 1976

Feurer, Rosemary, 'The meaning of "sisterhood": the British women's movement and protective labor legislation, 1870–1900', *Victorian Studies* xxxi (1988), 233–60

Frader, Laura L. and Sonya O. Rose (eds), *Gender and class in modern Europe*, Ithaca 1996

Fuchs, Rachel G., 'The right to life: Paul Strauss and the politics of motherhood', in Accampo, Fuchs and Stewart, *Gender and the politics of social reform*, 82–105

Gallagher, Catherine, 'The body versus the social body in the works of Thomas Malthus and Henry Mayhew', in Catherine Gallagher and Thomas Laqueur (eds), *The making of the modern body: sexuality and science in the nineteenth century*, Berkeley 1987, 83–106

Goodbody, John, 'The *Star*: its role in the rise of the new journalism', in Weiner, *Papers for the millions*, 143–63.

Goode, Erich and Nachman Ben-Yahuda, *Moral panics: the social construction of deviance*, New York 1994

Gray, Robert, *The factory question in industrial England, 1830–1860*, Cambridge 1996

Hall, Catherine, *White, male, and middle class: explorations in feminism and history*, New York 1992

Harrison, Barbara, ' "Some of them gets lead poisoned": occupational lead exposure in women, 1880–1914', *Social History of Medicine* ii (1989), 171–95

—— 'Women's health or social control? The role of the medical profession in relation to factory legislation in late nineteenth-century Britain', *Sociology of Health and Illness* xiii (1991), 469–91

—— 'The politics of occupational ill-health in late nineteenth-century Britain: the case of the match industry', *Sociology of Health and Illness* xvii (1995), 20–41

—— *Not only the 'dangerous trades': women's work and health in Britain, 1880–1914*, London 1996

Havighurst, Alfred F., *Radical journalist: H. W. Massingham, 1860–1940*, London 1974

Hepler, Alison, *Women in labor: mothers, medicine, and occupational health in the United States, 1890–1980*, Columbus 2000

Hewitt, Margaret, *Victorian wives and mothers*, London 1958

Holdsworth, Clare, 'Dr John Thomas Arlidge and Victorian occupational medicine', *Medical History* xlii (1998), 458–75

Humphries, Jane, 'Protective legislation, the capitalist state, and working-class men: the case of the 1842 Mines Regulation Act', *Feminist Review* vii (1980), 1–33

—— ' ". . . The most free form of objection . . .": the sexual division of labour and women's work in nineteenth-century England', *Journal of Economic History* xlvii (1987), 929–48

Jenson, Jane, 'Gender and reproduction: or, babies and the state', *Studies in Political Economy* xx (1986), 9–46

John, Angela V., *By the sweat of their brow: women workers at the Victorian coal mines*, London 1980

—— (ed.), *Unequal opportunities: women's employment in England, 1800–1918*, Oxford 1986

Jones, Gareth Stedman, *Outcast London: a study in the relationship between the classes in Victorian society*, New York 1971

Jones, Helen, 'Women health workers: the case of the first women factory inspectors in Britain', *Social History of Medicine* i (1988), 165–81

Jordanova, Ludmilla, *Sexual visions: images of gender in science and medicine between the eighteenth and twentieth centuries*, Madison 1989

Kent, Susan Kingsley, *Sex and suffrage in Britain, 1860–1914*, Princeton 1987

Knight, Patricia, 'Women and abortion in Victorian and Edwardian England', *History Workshop* iv (1977), 57–69

Koss, Stephen, *The rise and fall of the political press in Britain: the nineteenth century*, Chapel Hill 1981

Koven, Seth and Sonya Michel (eds), *Mothers of a new world: maternalist politics and the origins of welfare states*, New York 1993

Laqueur, Thomas, *Making sex: body and gender from the Greeks to Freud*, Cambridge, Mass. 1990

—— 'Sex and desire in the industrial revolution', in Patrick O'Brien and Roland Quinault (eds), *The industrial revolution and British society*, New York 1993, 100–23

Lee, Alan J., *The origins of the popular press, 1855–1914*, London 1976

Lee, W. R., 'The emergence of occupational medicine in Victorian times', *British Journal of Industrial Medicine* xxx (1973), 118–24

Levine, Phillipa, 'Consistent contradictions: prostitution and protective labour legislation in nineteenth-century England', *Social History* xix (1994), 17–35

Lewenhak, Sheila, *Women and trade unions: an outline history of women in the British trade union movement*, London 1977

Lewis, Jane (ed.), *The politics of motherhood: child and maternal welfare in England, 1900–1939*, London 1980

—— *Women in England, 1870–1950: sexual division and social change*, Bloomington 1984

—— 'The working-class wife and mother and state intervention, 1870–1918', in Lewis, *Labour and love*, 99–120

—— (ed.), *Labour and love: women's experiences of home and family, 1850–1914*, Oxford 1986

Liddington, Jill and Jill Norris, *One hand tied behind us: the rise of the women's suffrage movement*, London 1978

McFeeley, Mary Drake, *Lady inspectors: the campaign for a better workplace, 1893–1921*, Athens, Ga. 1991

Malcolmson, Patricia, *English laundresses: a social history, 1850–1930*, Urbana 1986

Malone, Carolyn, 'The gendering of dangerous trades: government regulation of women's work in the white lead trade in England, 1892–1898', *Journal of Women's History* viii (1996), 15–35
—— 'Gendered discourse and the making of protective labor legislation in England, 1830–1914', *Journal of British Studies* lxxiii (1998), 166–91
—— 'Sensational stories, endangered bodies: women's work and the new journalism in England in the 1890s', *Albion* xxxi (1999), 49–71
Mappen, Ellen, *Helping women at work: the Women's Industrial Council, 1889–1914*, London 1985
—— 'Strategies for change: social feminist approaches to the problems of women's work', in John, *Unequal opportunities*, 235–59
Mason, Michael, *The making of Victorian sexuality*, New York 1994
Meiklejohn, A., *The life, work and times of Charles Turner Thackrah, surgeon and apothecary of Leeds (1795–1833)*, London 1957
Michel, Sonya and Seth Koven, 'Womanly duties: maternalist policies and the origins of welfare states in France, Germany, Great Britain, and the United States, 1880–1920', *American Historical Review* lxxxxv (1990), 1076–108
Middleton, Lucy (ed.), *Women in the Labour movement: the British experience*, London 1977
Morris, Jenny, *Women workers and the sweated trades: the origin of minimum wage legislation*, Aldershot 1986
Mort, Frank, *Dangerous sexualities: medico-moral politics in England since 1830*, New York 1987
Moscucci, Ornella, *The science of woman: gynaecology and gender in England, 1800–1929*, Cambridge 1993 edn
Mosedale, Susan Sleeth, 'Science corrupted: Victorian biologists consider "the woman question" ', *Journal of the History of Biology* xi (1978), 1–56
Offen, Karen, 'Depopulation, nationalism, and feminism in *fin-de-siècle* France', *American Historical Review* lxxxix (1984), 648–76
Olcott, Theresa, 'Dead centre: the women's trade union movement in London, 1874–1914', *The London Journal* ii (1976), 34–50
Oppenheim, Janet, *Shattered nerves: doctors, patients, and depression in Victorian England*, Oxford 1991
Poovey, Mary, *Uneven developments: the ideological work of gender in mid-Victorian England*, Chicago 1988
—— *Making a social body: British cultural formation*, Chicago 1995
Posner, E., 'John Thomas Arlidge and the Potteries', *British Journal of Industrial Medicine* xxx (1973), 266–70
Rendall, Jane, *The origins of modern feminism: women in Britain, France, and the United States 1780–1860*, London 1985
Rose, Sonya O., 'Gender antagonism and class conflict: strategies of male trade unionists in nineteenth century Britain', *Social History* xiii (1988), 191–208
—— ' "From behind the women's petticoats": the movement for a legislated nine hour day and state protection of working women in Britain, 1870–1878', *Journal of Historical Sociology* iv (1991), 32–51

───── Limited livelihoods: gender and class in nineteenth-century England, Berkeley 1992

Ross, Ellen, Love and toil: motherhood in outcast London, 1870–1918, Oxford 1993

Rowan, C., 'Women in the Labour Party', Feminist Review xii (1982), 74–91

Rowe, D.J., Lead manufacturing in Britain: a history, London 1983

Russett, Cynthia Eagle, Sexual science: the Victorian construction of womanhood, Cambridge, Mass. 1989

Satre, Lowell J., 'After the match girls' strike: Bryant and May in the 1890s', Victorian Studies xxvi (1982), 7–31

Schiebinger, Londa, The mind has no sex? Women in the origins of modern science, Cambridge, Mass. 1991

Schmiechen, James A., Sweated industries and sweated labor: the London clothing trades, 1860–1914, Urbana 1984

Scott, Joan, Gender and the politics of history, New York 1988

Searle, G. R., The quest for national efficiency: a study in British politics and political thought, 1899–1914, Berkeley 1971

Seccombe, Wally, 'Patriarchy stabilised: the construction of the male breadwinner wage norm in nineteenth-century Britain', Social History ii (1986), 53–76

Soldon, Norbert, Women in British trade unions, 1874–1976, Dublin 1978

Stewart, Mary Lynn, Women, work, and the French state: labour protection and social patriarchy, 1879–1919, Montreal 1989

Stone, Judith F., 'Republican ideology, gender and class: France 1860s–1914', in Frader and Rose, Gender and class, 238–59

Thom, Deborah, 'The bundle of sticks: women, trade unionists, and collective organization before 1918', in John, Unequal opportunities, 261–89

Tosh, John, 'What should historians do with masculinity? Reflections on nineteenth-century Britain', History Workshop xxxviii (1994), 179–202

Tuana, Nancy, The less noble sex: scientific, religious, and philosophical conceptions of women's nature, Bloomington 1993

Valenze, Deborah, The first industrial woman, New York 1995

Valverde, Marianna, ' "Giving the female a domestic turn": the social, legal and moral regulation of women's work in British cotton mills, 1820–1850', Journal of Social History iv (1988), 619–34

Vernon, James, Politics and the people: a study in English political culture, c. 1815–1867, Cambridge 1993

Walkowitz, Judith, Prostitution and Victorian society: women, class, and the state, Cambridge, 1980

───── City of dreadful delight: narratives of sexual danger in late Victorian England, Chicago 1992

Warburton, W. H., The history of trade union organisation in the North Staffordshire Potteries, London 1931

Weaver, Steward Angas, John Fielden and the politics of popular radicalism, 1832–1847, Oxford 1987

Weindling, Paul, 'Linking self-help and medical science: the social history of occupational health', in Weindling, *Occupational health*, 1–31

────── *Health, race, and German politics between national unification and Nazism, 1870–1945*, Cambridge 1989

────── (ed.), *The social history of occupational health*, London 1985

Weiner, Joel H., 'How new was the new journalism?', in Weiner, *Papers for the millions*, 47–71.

────── (ed.), *Papers for the millions: the new journalism in Britain, 1850s to 1914*, Westport 1988

Whipp, Richard, 'Work and social consciousness: the British potters in the early twentieth century', *Past and Present* cxix (1988), 132–57

────── 'Kinship, labour and enterprise: the Staffordshire pottery industry, 1890–1920', in Pat Hudson and W. R. Lee (eds), *Women's work and the family in historical perspective*, Manchester 1991, 184–203

Wikander, Ulla, Alice Kessler-Harris and Jane Lewis (eds), *Protecting women: labor legislation in Europe, the United States, and Australia, 1880–1920*, Urbana 1995

Wohl, Anthony, *Endangered lives: public health in Victorian Britain*, London 1983

Unpublished paper

Hall, Robert G., 'Unsexing the male: gender, technology, the state, and Chartism in the cotton district, 1830–1860', paper presented at the 1993 annual meeting of the Southern Conference on British Studies

Index

Abraham (later Tennant), May, 33, 35–6, 43, 142
acrobats, 131
Acton, Dr William, 18
Alderson, Dr James, 98–9, 100
Anderson, Adelaide, 105–6, 116, 141
arbitration, pottery trade (1901–3), 65–8, 70
Arlidge, Dr John T., 53–4, 55–6, 57, 58, 67, 71, 95, 98, 103, 118, 119
Asquith, Herbert, 33–4, 39–41, 42, 48, 50, 141

Ballantyne, Dr J. W., 108–9
The barmaid problem, 125
barmaids, 7, 121, 124–32
Barmaids' Political Defence League, 128–31
Baron, Ava, 2, 10
Besant, Annie, 77–9
Blackburn, Helen, 120, 121, 138
Boer War, 143
Bondfield, Margaret, 123
Boucherett, Jessie, 16, 17, 24–5, 29, 31, 32, 42–3, 44, 64, 120, 121, 123
Bourgeois, Leon, 145, 147
Bridges, Dr J. H., 15
British Medical Journal, 38–9, 83–4, 100, 105
Bryant and May, 77–8, 79–82, 91
Burns, John, 40–1, 61, 62, 141–2
Butler, Josephine, 16, 17

Carlile, Richard, 13
Chadwick, Edwin, 12
coal industry, 5, 7, 25, 121, 132–6
Collett, Clara, 55, 56
Common Cause, 133–4
Contagious Diseases Acts, 16
cotton industry, 11–12, 13, 14, 15–16, 114, 128, 130, 136
County Advertiser for Staffordshire and Worcestershire, 25
Cramp, Dawkins, 55, 56–7

Daily Chronicle, 1, 74, 79, 93, 139–40; pit-brow work, 135; nail and chain trade, 27; pottery trade, 55, 57, 61, 65, 85–6, 87–9, 90–1; white lead trade, 4, 33, 36–9, 46, 53, 84, 100
Dangerous Performance (Women) Bill (1906), 131
Dangerous trades: the historical, social, and legal aspects of industrial occupations as affecting health, by a number of experts, 101–2
Deane, Lucy, 59–60
Deegan, John, 12
Delevingne, Malcolm J., 29
Delves, Frank, 46
Dickensen, Sarah, 128–30
Dietrich, Eduard, 148
Digby, Kenhelm, 64–5
Dilke, Charles, 61, 62, 87
Dilke, Emilia, 31, 47–8
Doherty, John, 11
Dunraven, Lord, 20, 23
Dyhouse, Carol, 113–14

Edwards, Thomas, 54–5, 67–8, 69
Engels, Friedrich, 11
Engerand, Fernand, 147
Eugenics Education Society, 9, 111
The evolution of sex, 113
Expectant motherhood: its supervision and hygiene, 108–9

factory acts: (1847), 1, 3, 10, 11; (1864), 54; (1874), 1, 3, 10, 14; (1891), 1, 3, 10, 21, 29–30, 58, 120, 139; (1895), 1, 4, 5, 10, 33–4, 41–50, 139, 141
Factory Acts Reform Association, 14, 16
factory inspectors, 3, 5–6, 30, 33, 99; pottery trade, 55, 56–7, 59–60, 61–2, 65, 105–6; white lead trade, 39–40, 45–6, 50–1
Fawcett, Millicent Garrett, 24, 26, 133–4
Fellows, Joseph, 28
feminists: barmaids, 7, 124–32; dangerous trades clause (1895), 41–8; equal-rights feminists, 6–7, 32, 136, 140, 141, 150; florists, 131–2; nail and chain work, 24–5, 30–2; pit-brow work, 7, 132–4, 136; pottery work, 64; protective

labour legislation, 6–7, 16–17, 120–4, 136–8, 140; social feminists, 6–7, 31–2, 122–4, 136–8, 140
florists, 131–2
France: protective labour legislation, 139, 144–7, 150
Frankland, Dr Percy, 104–5, 142–3
Freedom of Labour Defence Association, 6, 121–2, 128, 129, 138, 140

Gaskell, Peter, 9, 11, 139
Geddes, Patrick, 113
Germany: protective labour legislation, 139, 144–5, 148–50
Gladstone, Herbert, 130–1
Glasgow, 124–5, 131
Goadby, Dr Kenneth, 116–18, 119
Gooch, J. P., 124–5
Gore-Booth, Eva, 121, 124, 128–30, 131–3, 136–8
Gould, Edward, 50
government investigations: pottery trade (1860s), 53–4, 98; (1893), 56–7, 59; (1897), 59–60; (1898), 63–5, 101, 104–5; (1908), 68–9, 72, 106–8; white lead trade (1892), 35–6; (1893) 39–40, 41, 47, 48–9, 50–1, 100–1
Green, George, 23
Greenhow, Dr Edward, 17, 53–4, 98

Hamilton, Dr Alice, 116, 119
Hare, Edith, 43–4
Harmood-Banner, Sir J. S., 134
Harrison, Barbara, 96
Heather Bigg, Ada, 25, 29, 31, 121
Henderson, Arthur (factory inspector), 39
Henderson, Arthur (MP), 125
Hingley, B., 25
Holmes, Dr T., 15
Home Office, 1, 3, 4, 95, 135; barmaids, 125, 130–1; pit-brow women, 134–6; Factory Act (1895), 41–3, 45–6, 48, 49, 120, 141; nail and chain trade, 20–1, 25–6, 27–30, 31, 32; pottery trade, 58, 61–5, 70; white lead trade, 33–4
Homer, Thomas, 22, 27–8
House of Lords Select Committee on the Sweating System (1888–90), 3, 19, 20, 22–3, 30, 32

Industrial hygiene and medicine, 119
industrial illness and disease: lead poisoning, 1–2, 4–5, 33–45, 52, 53, 54, 56–7, 58–61, 63–8, 70, 72, 84–93, 96–106, 108–12, 114–20, 122, 138, 140; men's health, 66–7, 92–3, 115–16, 118–19, 122, 138, 140; phosphorous poisoning, 78, 79–82, 83–4, 90–1; pulmonary diseases, 53–4, 98, 99, 106
Industrial poisons in the United States, 116, 119
infant mortality, 1, 92–3, 99, 107–8, 114, 141–2, 143–4; cotton textile trade, 15–16, 17–18; pottery trade, 53, 59–60, 63, 65, 68, 72, 87–9, 93–4, 103, 105–6, 107, 109, 110, 112, 117, 123; white lead trade, 33, 36, 37–8, 40, 47, 51, 91, 100, 119, 120, 141
Infant mortality: a social problem, 108
Irwin, Margaret H., 43

Johnson, Richard, 50
Joint Committee on the Employment of Barmaids, 125–7, 128, 129
Juggins, Richard, 22

Kay, James, 12
King-May, Kate, 129, 133

Lancashire and Cheshire Women's Textile and Other Workers' Representation Committee, 124, 128, 133, 136
The Lancet, 98–9, 126–7
Laqueur, Thomas, 6, 12, 112
Legge, Dr Thomas, 5–6, 65, 66–7, 95, 102–4, 106, 116–18, 119
Leonard, Thomas, 13
Link, 77–9
Llewellyn, Arthur, 63
London, 19, 77–8, 80–2, 129, 132, 134
Lovatt, Joseph, 68
Lushington, Sir Godfrey, 28–9, 45

MacArthur, Mary, 123
MacDonald, Margaret, 121, 122–3, 125, 128, 131, 137, 140
McFeeley, Mary Drake, 60, 65, 94
McNish, James, 11–12
Malthus, Thomas, 10, 12
Manchester, 11, 16, 124, 132, 133
Manchester Guardian, 89, 128, 135
March-Phillipps, Evelyn, 46–7
Markham, Sir Arthur, 132, 136
Martin, Rudolf, 149
masculinity and work, 13, 66–8, 69–71, 92–3, 115–16, 118, 137, 140
Massingham, Henry W., 93

INDEX

match trade, 5, 77–8, 79–83, 90–1, 92
Matthews, Henry, 23, 25–6, 27–8, 29, 49
Maudsley, Thomas, 9, 15–16
medical profession, 112–13, 115; certifying surgeons, 5–6, 66–7, 99, 102–3, 117; and dangerous trades regulations, 2, 5–6, 40, 53, 63, 95–6, 99, 100–1; and future of the race, 6, 9, 97, 108–12, 119, 126–7, 142–3; lead poisoning, 2, 6, 37–8, 40, 51, 55–6, 58–9, 63, 66–7, 72–3, 108, 110–12, 114–19, 138, 140; medical officers of health, 6, 17, 99, 105–7, 109–10, 142; men's health, 53–4, 55–6, 66–7, 97–8, 101–3, 115–17, 118, 119, 140; opinions about women's work, 6, 9, 15, 17–18, 40, 43–4, 55–6, 66, 104–12, 113–15, 126–7, 134–5, 140; phosphorous poisoning, 81–2, 82–3; pulmonary diseases, 53–4, 56, 98; and women's greater susceptibility to lead poisoning, 6, 36, 37, 38–9, 40, 51, 53, 56, 58–9, 63, 72, 96–7, 99–104, 104–5, 106, 118–19, 140
Mort, Frank, 5, 115
Mundella, A. J., 14

nail and chain trade: employers and women's work, 23–4, 28; employment of women, 21–2, 25–6; feminists, 21, 24–5, 30–1, 32; Home Office, 20–1, 25–6, 27–30, 31, 32; House of Lords Select Committee on the Sweating System (1888–90), 3, 19, 20, 22–3, 30, 32; male trade unions, 4, 21–2, 27–8, 30, 32; newspaper articles about, 5, 21, 26–7, 30, 31, 74; proposed regulation of, 3, 20, 21–2, 29–30; sweated labour, 3, 19, 22, 32; women's work and reproduction, 3, 20, 21, 22, 26, 30
Nash, Vaughan, 93
National Industrial and Professional Women's Suffrage Society, 124
National Union of Women Workers, 127
National Union of Women's Suffrage Societies, 133–4
Newcastle-upon-Tyne, 34–5, 36–8, 39–43, 44–5, 50–1, 98–9, 100–2, 103, 111–12, 116–17
new journalism, 5, 31, 33, 53, 60–1, 74–9, 83–4, 90–1, 92–3, 95, 135–6, 139–40
Newman, Dr George, 108

O'Connor, Thomas P., 74, 76–7
Ogle Moore, Helen, 24–5, 29, 31, 43–4
Oliver, Dr Thomas, 95, 97, 104; and future of the race, 6, 9, 106, 110–12, 119; lead work and infant mortality, 36, 38, 40, 51, 88, 99, 102, 106; lead work and male reproduction, 115–16; opinions about women's work, 111–12, 114–15, 137, 140; pottery report (1899), 63–6, 101, 105, 142; and women's greater susceptibility to lead poisoning, 6, 36, 37, 38–9, 40, 51, 58–9, 63, 65, 99, 101–2, 106, 118, 140; service on White Lead Committee, 40, 101
Oram, R. E. Sprague, 33, 39, 45, 50
Orme, Eliza, 43, 125
Owen, William, 54, 55

Pall Mall Gazette, 26–7, 30, 76, 77, 93, 135
Paterson, Mary, 59–60
Patten, Dr Cooper, 109, 110
Penny Illustrated Paper, 86, 87
Pic, Paul, 147
Pilling, Richard, 12
The potter and lead poisoning, 88, 105
Pottery Gazette, 57–8, 89–90
The pottery manufacture in its sanitary aspects, 55–6, 57
pottery manufacturers, 53, 57–8, 63–4, 70, 72, 89–90
pottery trade, 1, 3, 51; arbitration (1901–3), 65–8, 70; employment of women, 52, 69, 71–2; designation as a dangerous trade, 52; government investigation of (1860s), 53–4, 98; (1893), 56–7, 59; (1897), 59–60; (1898), 63–5, 101, 104–5; (1908), 68–9, 72, 106–8; Home Office, 58, 61–5, 70; lead poisoning, 51, 54, 56, 98, 99, 103, 117–18; male trade unions, 54–5, 67–9; men's health, 53–4, 55–7, 66–7, 98, 103, 117, 118, 140; pulmonary disease, 53–4, 56, 98; newspaper articles about, 1, 53, 55, 60–1, 65, 79, 84–90, 91, 92, 93–4, 140; special rules (1894), 52, 58; (1898), 5, 52, 63, 66, 71, 139; (1899), 52, 66; (1903), 52, 71; (1913), 52, 72; trade unionism among women, 48, 62; women's work and infant mortality, 53, 59–60, 63, 65, 68, 72, 87–9, 93–4, 103, 105–6, 107, 109, 110, 112, 117, 123
Prendergast, Dr W. Dowling, 88, 105, 106, 119

Price, William, 27, 30
Public Health Act (1872), 6

Reddish, Sarah, 129
Reid, Dr George, 106–7, 123
Report on infant mortality in Lancashire, 114
Ridley, Matthew, 62–3, 64–5, 85–7, 87, 91
Robinson, Brooke, 25
Roper, Esther, 121, 124, 128–9, 131–2, 136
Rose, Sonya O., 14–15
Royal Commission on Labour (1892–4), 33, 35–6, 43, 54–5, 67, 125
Russell, George, 43
Rylett, Rev Harold, 22

St James's Gazette, 86
Scott, Joan, 1, 10
Shaw, George Bernard, 93
Sheffield, 114–15
Simon, Dr John, 53, 54, 98
Smith, Herbert Llewellyn, 93
Society for Promoting the Employment of Women, 6, 16, 24–5, 31, 32, 43–4, 48, 64, 120, 122, 136, 140
special rules: match trade (1898), 82–3; pottery trade (1894), 52, 58; (1898), 5, 52, 63, 66, 71, 139; (1899), 52, 66; (1903), 52, 71; (1913), 52, 72; white lead trade (1892), 33, 38, 47–8; (1898), 34, 50, 93, 111–12, 139
Spencer, Herbert, 113
Staffordshire, 21, 23–31, 48, 52–3, 55–72, 84–5, 87–93, 98–9, 101, 104–8, 110, 112, 117–18
Staffordshire Sentinel, 26, 84, 92
Stanhope, Paul, 20, 29
Star, 1, 27, 30, 74, 76–7, 79–84, 86, 88, 90, 91, 93, 125
Statement of an amateur pitbrow worker, 133
Stead, William T., 5, 74–5, 76, 139–40
Stoke-on-Trent, 52–3, 55
Strauss, Paul, 146–7
Stuart-Wortley, C. B., 49

Tennant, Henry J., 49, 62
Tennant, May, 142
Thackrah, Charles Turner, 97–8
Thompson, J. Arthur, 113
Thorpe, T. E., 63–5, 66, 104
Thring, Lord, 20, 23
The Times, 24, 125
Trades Union Congress, 4, 19, 22, 31, 46

trades unions, 4, 15–16, 19, 21–2, 27–9, 54–5, 62, 68–71, 77, 123, 136
Troup, C. E., 61, 62
Tuckwell, Gertrude, 31, 46, 64, 120–1, 123

Vaughan, A. P., 50
Veness, Annie, 45
Vigilance Association for the Defence of Personal Rights, 16–17, 150
Vynne, Nora, 121, 138

Walkowitz, Judith, 5, 78–9, 90–1
Webb, Sydney, 93
Westminster Gazette, 86–7
White Lead Act (1883), 33
White lead committee (1893), 39–40, 41, 47, 48–9, 50–1, 100–1
white lead trade, 1, 64; designation as a dangerous trade, 33; employment of women, 34–5; government investigation of (1892), 35–6; (1893), 39–40, 41, 47, 48–9, 50–1, 100–1; Home Office, 33–4; lead poisoning, 35, 50–1, 97–9, 100–3, 116–17; men's health, 35, 50–1, 97–8, 101–2, 115–17, 140; newspaper articles about, 1, 4, 33, 36–40, 46, 140; special rules (1892), 33, 38, 47–8; (1898), 34, 50, 93, 111–12, 139; white lead committee, 39–40, 41, 47, 48–9, 50, 100–1; women's wages in, 34–5; women's work and infant mortality, 33, 36, 37–8, 40, 47, 51, 91, 100, 119, 120, 141
Whitelegge, Dr Benjamin Arthur, 5, 46, 61, 65, 70, 101, 131
Whyte, Eleanor, 42
Wigan, 132, 133–5
women: foetal protection, 3, 7, 10, 34, 53, 97, 110, 138–43, 150; maternity, 10, 15–18, 107–8, 111, 122–3, 126–7, 137, 138, 139–44, 150; reproductive health, 1–4, 20, 22–3, 26–7, 30–1, 35–8, 40, 46–7, 54, 58–61, 63, 65, 72–3, 87–9, 91, 93–4, 96, 100, 103–11, 116, 119–20, 135, 137–8, 139–40, 141–3; sexuality, 1, 3, 9–10, 13–14, 18, 22–3, 92, 126, 139; suffrage and protective labour legislation, 124, 129–30, 132–3, 136; sweated labour, 3, 19, 21–30, 77–9, 90; wages and conditions of work, 22–6, 34–5, 36–7, 50–1, 52, 54, 69, 78, 130, 132–3; women's work and future of the race, 7, 9, 97, 108–11, 119, 126–7, 139–44, 150

Women as barmaids, 125, 129
women factory inspectors, 59–60, 61–2, 103, 105–6, 116, 141
Women's Cooperative Guild, 46–7, 89, 120, 133
Women's Emancipation Union, 44–5, 141
Women's Employment Association, 45
Women's Freedom League, 134
Women's Industrial Council, 120, 127, 131, 134
Women's Industrial Defence Committee, 42, 44, 48, 120, 122, 136, 138, 140, 141
Women's Labour League, 6, 123–4, 137, 140
Women's Liberal Federation, 44–5, 89, 127
Women's right to work, 137
Women's Trade Union League, 4, 6, 31–2, 47–8, 62, 64, 131–2, 140
Woods, Samuel, 134
Wynne, F., 135–6
Workmen's Compensation Act (1897), 62, 70, 85